take away **Energy**

take away **Freedom**

Know, Understand, Influence

The role of energy in your life &
how environmentalists control its use

Marita Littauer Noon

IPG

Intermedia Publishing Group

Energy Freedom

Published by:
Intermedia Publishing, Inc.
P.O. Box 2825
Peoria, Arizona 85380
www.intermediapub.com

ISBN 978-1-937654-05-4

Printed in the United States of America

Dedication

To all my mentors.
Without your investment in me,
this book would not exist,
especially:

Florence Littauer, my mother,
thanks for teaching me the power of words.

Ron Arnold,
thanks for believing in me and pouring your knowledge into me.

John Clema,
thanks for making me connect the dots every day.

John Fund,
thanks for opening doors for me.

Michael Economides,
thanks for answering my endless energy questions.

Contents

Introduction

Green Is Good, Isn't It?

"Greed is good" was far more than a catch phrase made part of the vernacular through the popular late '80s movie *Wall Street*. As is often the case, art imitates life. "Greed is good" was more than a line from a movie; it was the mantra of the times. However, as Gordon Gecko, the famed character played by Michael Douglas, found out, too much greed was not good. His tactics sent him away for twenty years.

Today, in an age of simplicity and even austerity, "greed is good" sounds horribly outdated. Replacing it would be the slogan, "green is good."

Anything or anyone who can label one's self as "green" has a perceived marketing advantage. Without fully understanding the implications of "green," people support the "green" concept as generally being better for the environment. Without a specific universal definition of "green," products as diverse as political candidates, diapers, and cars proudly sport the moniker. Though like Gordon Gecko learned, it is possible to have too much. Blind adoption of concepts labeled "green" may imprison America for the next twenty years or more. Resources, which America needs to move forward, to balance out our trade deficit, and to remain a super power, are being locked up.

The resources I am referring to are energy, specifically oil, gas, coal, and uranium—the essential ingredients that fuel America. But other natural resources are also a part of what supplies our energy. We need copper for wires in transmission lines. Sulfur, lead, lithium, and rare earths are needed for batteries—and these resources, too, are being locked up. While most people only think of gasoline and/or electricity when they think of energy, if they think about it all, there is much more involved.

Wyoming's new Governor Matt Mead tells of a man he met while campaigning. The man told him in no uncertain terms that the country needs to stop using "dirty coal." Governor Mead asked the man what he proposed we use instead. "Electricity!" was the man's proud response—not realizing electricity comes from coal.

Most people do not give energy a thought, and, like the man Governor Mead encountered, they probably think electricity comes from the wall. Despite people's low cognizance, energy is responsible for the safe and comfortable lives we lead.

Shortly after I first became aware of the role energy plays in modern life, I went on vacation to Grand Cayman. Preparing for the trip, I'd read that virtually nothing is grown on the island and almost everything is shipped in. It made me wonder about the island's source of power. I began asking the service people I encountered about their electricity. I asked the hotel manager, cab drivers, and waitresses, "Where does your electricity come from on this island?" Not one had an answer. I was there five days before a bartender summoned the manager, who told me, "Diesel-powered generators." The diesel fuel was shipped in, probably in diesel-powered ships.

The Grand Cayman quest reminded me of my honeymoon in the British Virgin Islands. We were on Virgin Gorda when I ordered a piña colada. With his heavy accent, the bartender told me, "We have no courant." At least that is what I thought he said. It took a few tries before I understood that he'd said they had no "current." The power was out!

While most of us do not give energy a thought until "we have no current," it is the first thing we want restored following a natural disaster.

In the wake of the 2008 gasoline price spike, Senate Leader Harry Reed claimed: "Oil and gas are making us sick," when in fact they are what keep us well. Energy saves lives. When fire strikes or hurricanes are bearing down upon a city, it is energy—in this case in the form of gasoline—that allows people to drive away and escape death. One of the biggest hindrances to rescue efforts following Japan's earthquake/tsunami crisis was the lack of gasoline. When weather is extreme, it is energy—usually in the form of electricity (most frequently from coal or natural gas)—that keeps people alive. Air conditioning allows people to live in comfort in Arizona in the summer. Heating keeps people from freezing to death in Alaska in the winter.

"Britain is Freezing to Death"

Sunday Express, 5 December 2010

Tracey Boles

Middle-class families are among millions of Britons who cannot afford to heat their homes this winter, as elderly ride on buses all day to stay in the warm. After a week of snow and freezing temperatures, a shocking picture has emerged of the bleak months ahead for 5.5 million households.

Pensioners, who are among those most vulnerable to the cold, are resorting to extraordinary measures to keep warm.

Many have been using their free travel passes to spend the day riding on buses while others are seeking refuge from the cold in libraries and shopping centres.

Dot Gibson, spokeswoman for pressure group the National Pensioners' Convention, said: "Now that we have one of the coldest winters, older people are going to have to make the unenviable decision whether or not to put the heating on. The Government should guarantee that they won't cut the winter fuel allowance."

The death toll from the big freeze rose to seven yesterday. They included two men who were killed in a crash on the M62 in Humberside and two teenage girls who died when their car collided with a Royal Mail van in Cumbria.

The winter death toll is set to rise steeply as official figures show that nine elderly people died every hour because of cold-related illnesses last year. The number of deaths linked to cold over the four months of last winter reached nearly 28,000.

Charities claim this country has the highest winter death rate in northern Europe, worse than colder nations such as Finland and Sweden.

About half of the people forced to spend over 10 percent of their income on energy bills—the official definition of fuel poverty—are aged over 60.

But working families also face a tough time meeting the cost of keeping the central heating turned on as fuel prices continue to rise.

Ann Robinson, director of consumer policy at price comparison service uswitch.com, said: "Middle-class households are now in fuel poverty."

National Energy Action estimates that 5.5 million households will have plunged into fuel poverty by early next year due to price rises.

This is up 400,000 on the group's last estimate and represents 21 per cent of the UK's 26 million households.

The last official figures, for 2008, showed there were 4.5 million fuel-poor households in the UK. On Friday, British Gas will raise prices for eight million customers. Millions more customers of Scottish & Southern Energy and Scottish Power have already been hit by price rises.

Last winter 70 percent of households were forced to cut down or ration their energy use because of cost.

Uswitch's Ms. Robinson, who advised Tony Blair's government on energy policy, warned: "Winter price hikes will simply force even more people down this route."

Energy keeps us well. Energy makes us comfortable—as the sidebar from the British *Sunday Express* demonstrates. While the stoic nature of the Brits enables them to endure a lot, the senior citizens of the United States will not tolerate similar treatment without a payback. In America, without our expected energy, we'd be "squealing like pigs caught in a gate."

> Electricity is the energy commodity that separates the developed countries from the rest. Countries that can provide cheap and reliable electric power to their citizens can grow their economies and create wealth. Those who can't, can't.
>
> Robert Bryce, ***Power Hungry***

Not only does energy keep us well and make us comfortable, energy makes America great! Really, energy makes England great, too, and energy is what makes the developed world, "developed." Look at the following chart. On the left are examples of things that we expect in America and in the developed world—things that we take for granted. On the right is a comparative status in developing countries. What is the common denominator of the left column vs. the right? Hmmm… could it be "energy" or the lack thereof?

America	**Developing Countries**
Bottled water	Want clean water
Food network	Need food
Argue over light bulbs	Live in the dark
Wi-fi is a daily expectation	"why not?" is a daily question
Search for "Cold beer"	Stuck with "it's cold here"
Tanning salons are taxed	No choice but to work in the hot sun

Costco will deliver a shed for $999	Live in a shed
Any book in 60 seconds, read in any size font	Can't read
Cooking gadgets like rice cookers	Only have a bowl of rice
Instant access to information	Exist without basic education
Push a lever, waste disappears	Collect dung as cooking fuel
Clean teeth with whirs, hisses and flavor choices	Don't have teeth
Treadmills and ellipticals to cut down on body mass	No fat on their bones to eliminate
Traffic jams and mass transit	Never travel far from home
Choose between bed composition and price	Happy just to have a bed
Hot shower, preferred water stream	Shower in a stream

When it comes to energy, advertising and the media tell us "green"—typically referring to wind, solar, and biomass—is the preferred way to go (even when not economically viable or beneficial). Despite the anointed status, many "green" projects get held up for a variety of reasons, as we'll address.

As an example, a 2010 poll done in New Mexico by Public Opinion Strategies indicated that, while the vast majority of respondents acknowledged that natural gas was "important" or "very important" to New Mexico's economy, more than one-third of those same people believed that green energy should be pursued rather than natural gas. "Green is good." This attitude, on the part of both the public and the politicians making the decisions, allows known and available resources to be locked up. This is not just a New Mexico problem; it is a national issue.

But like greed, could too much green be bad?

Public perception seems to be too much green is never enough; green campaigns deserve implementation. Green is the driver for American energy policy—and we're being taken for a ride.

What would the world look like if we could wave a magic wand and suddenly make the wishes of the environmental thought leaders come true as they relate to energy? Would green be good? Would it be the environmental utopia we might expect?

The answer requires researching what the environmental groups, such as the Sierra Club, the Theodore Roosevelt Conservation Partnership, and Trout Unlimited, are proposing—so I did. The results became the foundation for my passion—some would say "compulsion"—to talk about energy to everyone and anyone "willing to listen."

My intent for this book is to imbue you with energy awareness. As you acknowledge the role energy plays in your life—how it keeps you safe and warm, how it provides your food, and gives you freedom; and embrace the fact that "energy makes America great," you will fight for it; you will not want to live like those Brits suffering in "fuel poverty."

In my personal energy education, I've read many exhaustive tomes offering a thorough treatment on the subject. Engineers or professors wrote the books. If you want to understand the difference between a watt, horsepower, or a joule, I can recommend several books for you—but this is not that book. I am writing for the average American energy consumer who knows that energy costs are going up, but doesn't understand why. I am writing for the person who is newly politically engaged out of concern for the direction America is heading. This book is for the person who wants to know what time it is, but not how to build a watch.

In the following pages, you will find four sections. Each can be read alone—though I hope you will start at the beginning and read from front to back.

Section One is based on my original research. It is a fun, light read that will shed a different light on "green" from the one offered by Madison Avenue. In Chapters one through six, I draw the green goals out to their logical conclusion. If you are new to political policy or question the motives of the green movement, please be sure to read Section One. If you have been following environmental extremism for years and are already convinced that too much green is not good, you may want to skip Section One and get right to the meat.

Section Two explains who the greens are, where they came from, and how they got so much power. As I speak on energy, and specifically the influence of environmentalists, I am frequently asked, "Why do they do this?" I often hear: "They are destroying America." If you've asked this question or had the same thoughts, be sure to read Section Two.

Section Three offers an overview of several green schemes throughout history. Again, the topics presented are a sampling of campaigns that have changed an industry or damaged the economy for naught. This section will open your eyes and surely give you fodder for conversation.

Sections One through Three are about waking up. If you do not read Section Four, you are likely to be left feeling frustrated and helpless. Section Four is about showing up, standing up, and speaking up. In Section Four, you'll find an action plan that will help you make a difference!

Throughout all four sections, you will find sidebars and pull quotes. The sidebars are stand-alone articles or commentaries that add to the topic being discussed. Those without the name of the author are written by me—usually previously published in a newspaper or on a website (date of original publication included). You will find that because I live in New Mexico, more of the specific issues addressed are from New Mexico. However, none of the core concerns is unique. They have been included here because of their universal application to the American energy story.

Often the added features are from others. In these cases, their byline will be included. The pull quotes—the brief text set within a frame—are a concise comment that illustrates the concept being presented. Pull quotes are usually from a noted expert on the topic.

While this book is not written as a textbook, you will find that it is fully researched and documented. Sources are cited, and you can find out more about any topic that interests you.

I started the Environmental Utopia project by looking at both well-known environmental groups and smaller, regional, and/or specifically targeted-issue groups to draft a broad view of the environmentalists' goals as they relate to energy. My initial work was done in 2007 and was used primarily for my personal education. However, the poll data I received from Public Opinion Strategies regarding New Mexico's view of the need for green energy highlighted the disconnect between the "green is good" ideal and the "fuel poverty" reality. Since then, I've updated my study with fresh research, and I present it in a tongue-in-cheek narrative based on the facts my research revealed.

Do you think living in a world where the environmentalists were in charge would be the "environmental utopia" we'd expect? Would we have "rivers and streams running so clear and clean that you can bend a knee to the water, cup your hands, and drink without fear," as one website states as their goal? Would we be able to "explore, enjoy and protect wild places of the Earth,"[1] as another posits? Would environmental utopia feature a "rural character" and a "unique quality of life,"[2] as proposed by yet another?

Hardly.

By using the combined energy goals of the twenty-five different groups, I analyzed and applied them to four key areas: transportation, modern conveniences, health, and housing. It became clear that the world would look nothing like what we enjoy today.

Read on. You will see that like greed, too much green is not good. Yes, we all do want to take care of the earth and wisely use the resources it gives us, but just because something is labeled "green" does not mean that it is really the best for us today.

Freedom of choice is one of the great things about America. If someone wants to live "off the grid," or even in a cave—they are free to do so. But that should be a choice, not something that is regulated or legislated. Energy does make America great. A world without energy would hardly be an environmental utopia.

Do you want to live in Environmental Utopia? Let's see what it would be like.

Section One

Environmental Utopia

I Like Oil

Baxter Black, DVM[3]

I must confess I like oil and I like factory farming
And I think factory health care is absolutely charming
And factory education lets us school the huddled masses
While Conoco and off shore rigs produce our natural gasses.

Factory transportation is a middle seat in coach
It's soccer moms in minivans, the thrifty man's approach.
If there ain't no fancy Lear jet parked outside your bungalow
It still means you can pack your lunch and stand in line and go!

I love the pharmaceuticals who've given us new life
From cancer cures to botox shots, reduced our stress and strife
And doubled up our life span, though you hear them cursed a lot
By the very ones whose world they've changed, I guess they just forgot?

Forgot that all the coal they hate gave birth to industry
That heats our homes and gives us cars and opportunity
And jobs, and time to stop and rest, take respite from the toil.
We built a country, good and great, with blood and sweat and oil.

There's some who sneer and denigrate what we call factory farming
Though we can feed the multitudes, they find it most alarming
"They're using everything," they say, "to make it more efficient.
And cheaper for the common folks. Do they think that they're omniscient?

Folks, there's nothing wrong with gardens or organic, we all know
The problem is most can't afford food that costs so much to grow.

If the world was a perfect place, the "complaining" crowd's ideal
We'd have no impact on the earth, bananas with no peel,
Forests without clearcut tracts, factories without smoke,
No endless freeways, windmill farms, no global warming jokes.

But none of us, 'specially farmers like the crowding and pollution
But with the help of oil and gas, there came a revolution
Where people could find work to live, and improve their situation
And farmers who could feed themselves, were asked to feed the nation.

And we who till and mine and build your malls and schools unending
Leave footprints with each concrete pour that take millennia in mending
The search to give us all good lives, and all... not just a few,
Should also strive to make our tracks less harmful when we do.

So here we sit on God's green earth, each one a worker bee
Contributing our own small part to have prosperity
And I for one am thankful what oil and factory farming gave
For if we'd stopped each time they whined, we'd still live in a cave!

Chapter 1

Perfect Paradise

If you've been house hunting, you know reality seldom lives up to the hype. "Charming and cozy" really means small and cramped. "Great possibilities" tells you that it is a serious fixer upper. "Great use of space" equates to tiny.

Brochures for housing developments promise an eternity of beautiful sunsets and an enviable lifestyle. I am reminded of advertising for a new community just outside of Albuquerque. For months I've seen glossy magazine ads with pink clouds, soaring vistas and dirt roads. The simple text stated:

"Wish you were here."
—Your Imagination
And "Discover new things,
like the old you."

Based on the quality of the ads, I mistakenly presumed the subdivision was in the prestigious foothills of the Sandia Mountains.

Here is how the builders describe the master-planned community on their website: "...The concept behind the community is to perfectly meld state-of-the-art homes and community buildings with the natural beauty that is New Mexico. Thus, as you wander through the neighborhoods and trails at Mariposa you will experience the delicate balance of nature, architecture and community, ... So while residents are surrounded by unsurpassed natural beauty, the amenities of the city are only minutes away."

Recently, I discovered that Mariposa is located on the outer edge of civilization, carved out of barren land, previously filled with tumbleweeds. A place where, without weeds, the winds blow the sand of your neighbor's yard into yours. A closer look at the development's early ads, does show sagebrush and tumbleweeds—but they are presented so majestically, you don't think of them as mere weeds.

Back in the eighties, when I lived in Los Angeles, I remember looking at homes that were similarly advertised. The houses surrounded a golf course, were nestled in the hills, and had majestic views. Mountaingate was a beautiful setting, but it was built around a landfill topped with a golf course. That was never mentioned in the advertising. Today, the homes are a million dollars plus—I should have bought one. But the golf course requires extensive maintenance to overcome the shifting of the land.

Both Miraposa and Mountaingate are beautiful communities. But they are not the perfect paradise as advertised—yet somehow we all want to believe that such a place exists.

The following phrases could have come from a real estate brochure:

"…rivers and streams running so clear and clean that you can bend a knee to the water, cup your hands, and drink without fear."

"…explore, enjoy and protect wild places of the earth."[4]

"…protect all wildlife and plants in their natural communities."[5]

"…protection of the forests, rivers, deserts and grasslands."[6]

"…to protect our water and air, our public lands, our rural character, and our unique quality of life."[7]

I want to go there. Don't you? Who doesn't want to live in a place with clean water? Who wouldn't want to live in a natural community with rural character and a unique quality of life? In this utopia there would be wildlife, plants, forests, rivers, deserts, and grasslands.

But this text is not from a slick, stylized real estate brochure. The phrases are collected directly from the home pages of various environmental groups' websites.

While the environmental groups do not state this, I am sure this environmental utopia would also have cherubic children frolicking in fields of daisies while baby bunnies nibble on the fresh green grass moistened by summer showers. Gentle breezes would blow and the sun's rays would glisten off the sparkling stream as it gurgles over granite stones. Puffy clouds in the shapes of flowers, hearts, and even Mickey Mouse would float by.

We all want a green earth, clean water, fresh air, and a safe food supply. Wouldn't it be great if we could quit fighting all the evils in the world, suddenly wave a magic wand, and grant all of these environmental groups their wishes?[8] Wouldn't the world be a prettier place, a perfect paradise? Does it sound too good to be true?

It is.

Like the real estate verbiage, environmental groups are equally deceptive. To learn the truth about the master-planned communities, you have to read the fine print. To learn the truth about the goals of the environmental groups, you have to dig deeper than the home page as I've done and exposed on the following pages.

If they could have all their dreams come true, we'd revert back to, at the least, Colonial days and very possibly—as you'll see—we'd all be living in caves. In reality, I believe we'd all revolt before that happened. But here's what it would look like.

With approximately six billion people on the planet, assuming an average of four people per household, we'd need 1.5 billion caves—and those caves would all need to be near a fresh water source. With no plumbing or sewage processing, that water would not be fresh for very long. The hillsides surrounding these caves would be barren—stripped of wood for heating in the cold months. The critters you'd think live nearby, providing a food source, would have been over hunted, and those remaining would have fled for "greener pastures." In just a short time, people would be dirty, hungry, sick, and naked—not to mention broke.

How did we get from the perfect paradise to pandemonium with one wave of the wand? To find out, read the following carefully documented, educational, and, hopefully, even entertaining chapters.

Yes, you will see I have taken this idea to a bit of an extreme to make a point—to the logical conclusion, but it does expose the true goals of a variety of environmental groups. Like the layers of an onion, we will peel away the elements of American life that these groups would have us eliminate, if we could truly wave that magic wand and give them their wishes.

Much of what would disappear are the very same things that have made America great—a country people are giving their lives for, and a nation into which people are risking their lives to enter.

You will find that the energy that sustains, enriches, and improves our lives is under attack. I believe people are smart, and that if they understand the complete energy picture, they will make wise choices. Unfortunately, most environmental groups present only one side of the picture, while ignoring the very real impacts their policies would have. As a result, most people do not know the truth about their real motives or understand the vital role energy plays in America.

Once that truth is known, people will make more informed decisions regarding the politicians for whom they vote, the groups to which they donate, and the choices they make for their personal lives. The idealistic-sounding choices of some are creating hardships for many.

Let's visit the perfect paradise of the master-planned community known as Environmental Utopia.

Chapter 2

Transportation: Getting Around in Environmental Utopia

"You are now free to move about the country" is more than just the Southwest Airlines slogan. The reason it works for the airline is that it resonates with Americans. We expect the freedom to move about the country, to go where we want, when we want. Whether by plane, train or automobile, we Americans count on our transportation.

With gas prices rising and the burning of carbon being blamed for global warming, the use of gasoline and diesel fuel, in cars and other vehicles—and transportation in general—frequently makes the headlines. They are never far from the mind of the environmentalists either. Numerous groups—including the National Resource Defense Council, the Sierra Club, and the Center for Biological Diversity—are actively working to prevent drilling for oil and gas and to expand moratoriums on drilling.[9] Earth First is even opposed to ethanol[10] and many other organizations are now turning against the presumed wonder fuel. It is these groups who have been pushing the energy agenda.

Government shut down of America (15 July 2010)

With all the concern about jobs, spending, and government transparency, many Americans don't think working on protecting outdoor spaces or creating more national monuments is a priority. Yet, this is where our government has been focusing.

In April 2010, a meeting was held at the White House that got virtually no press attention. The literature stated the purpose as bringing "together leaders from communities across the country that are working to protect their outdoor spaces."

At that meeting, President Obama signed a Presidential Memorandum establishing the America's Great Outdoors Initiative to promote efforts to conserve outdoor spaces. The press release states: "The Initiative will support a 21st-Century conservation agenda that builds on successes in communities across the country, and will start a national dialogue about conservation that supports the efforts of private citizens and local communities." It sounds nice, something we can all feel good about in a time of negative news.

The Department of Interior (DOI) sent a team of senior representatives from the DOI, the U.S. Department of Agriculture, and the U.S. Environmental Protection Agency around the country for public listening sessions and discussions on conservation, recreation, and reconnecting people to the outdoors.

Just the cost of the tour should be enough to have the initiative on the House of Representatives cost-cutting "You-Cut" agenda, let alone the cost of the greater program. But this is about far more than foolish government spending, and it is more than a distraction from the major issues facing America. It is about a government land grab and blocking access to America's natural resources.

In late January 2010, an internal memo was leaked, stating the White House had plans to create additional national monuments. The list contains fourteen possible locations totaling more than 13 million acres. The sites "may be good candidates for National Monument designation and the Antiquities Act." Many of them—such as Otero Mesa in south-central New Mexico, Vermillion Basin, Colorado, and Montana's Northern Prairie—have known energy resources.

The America's Great Outdoors Initiative could be one more federal effort to lock up land and prevent productive activities such as farming, ranching, logging, and resource extraction from taking place.

As happened with Mount Taylor's Traditional Cultural Property Designation in New Mexico or the Upper Missouri River National Monument in Montana, the federal land grabs have the potential to limit or further restrict the use of private and public land.

Montana rancher, Matt Knox, said the ranchers worked out a viable management plan with the BLM after Clinton's eleventh-hour executive order created the National Monument in 2001. But now, environmental groups are suing the BLM in an effort to remove virtually all activity, including grazing, boating on the river—even roads. Knox explained that because the area contains a mix of federal and private lands that are managed together, it is not possible for ranchers to fence their cattle onto their private property. Ranchers in the area have been fighting over the management plan for ten years, and it looks like the fight will never end.

Mining is impacted as well. In Montana's Flint Ridge Inventoried Roadless Area, the Forest Service approved a "Plan of Operations" that in practice locks up resources. They will allow mining, if a team of mules is used to haul the equipment, hand tools are used to clear the land and reclaim it, and the mules are fed "certified weed-free hay."

When asked about additional national monuments in Montana, Secretary of the Interior Ken Salazar denied that he had plans for additional national monuments "unless there is significant public involvement, discussion and debate over any such proposal."

Can you say "listening session?"

Regarding the listening sessions, BLM Director Robert Abbey said, "Let's go out and open a dialogue with the American public and hear directly from them what participants consider to be important."

With this in mind, it is imperative that those of us who value private property, multiple-use lands, and recreation must attend the listening sessions to have our voices heard. This, plus contacting our Senators, is the forum afforded to us.

We must stop the government from shutting down America!

Lands with known natural resources have been locked up under the heading of "National Monument," "Traditional Cultural Property," "Wilderness and Roadless Area" or other "protection"-aimed designations. Government regulation and legislation are being used to prevent access. In other cases, the miles of red tape make accessing the resource nearly impossible. Instead of trying to create new wealth, the current tone in government is to do everything possible to block, kill or deny. The only thing that seems to be expanding is moratoriums.

Sen. Bingaman is making a big mistake with his dangerous wilderness bill

Washington Examiner, 26 November 2010

Ron Arnold

Sen. Jeff Bingaman is being called our worst border security threat. Angry Doña Ana County residents have branded the New Mexico Democrat's Organ Mountains-Desert Peaks Wilderness Act as "Bingaman's Bandit Boulevard" for proposing a 50-mile-long safe haven for Mexican drug runners—and worse.

John Hummer, former chairman of the Greater Las Cruces Chamber of Commerce, reminded his community of the no-motorized-vehicle clause in wilderness laws. The Border Patrol can't patrol.

Wilderness laws allow our own park and forest rangers to keep the cops out. That's supposed to protect nature, but ends up protecting drug cartels, illegal immigrants—and terrorists.

Steve Wilmeth, fifth-generation New Mexico rancher, told me that Bingaman's north-south strip and two mountain clusters don't worry him for the inevitable Mexican intruders, but for the OTMs—Other Than Mexicans.

A U.S. Border Patrol document obtained by *The Examiner* shows the nationality and number of OTMs arrested last year. A few samples: Afghanistan (12); Indonesia (95); Iran (42); Iraq (42); Jordan (52); Saudi Arabia (6); Somalia (70); Yemen (22).

Members of the National Association of Former Border Patrol Officers say they weren't smuggling drugs. Retired Border Patrol Officer Zack Taylor said Bingaman's wilderness boundary is just a stroll from violent Juarez and El Paso.

Terrorists want terror, said Taylor—unexpected targets with severed body parts and dead babies. A dozen relays of three sunburned hikers carrying big backpacks could trek unmolested up Bingaman's Boulevard, stockpiling materials to obliterate the balloon festival, the state Capitol, the Acoma Pueblo—anything we treasure.

Frank DuBois, a Reagan-era deputy assistant secretary of the interior, told me that interdicting criminals in wilderness is losing priority. Bingaman's bill makes his wilderness a "component of the National Landscape Conservation System." That's ominous.

In a previous Examiner report, I exposed Wendy Van Asselt's Wilderness Society campaign with Bureau of Land Management officials that led her to the House Natural Resources Committee, where she helped create that hands-off conservation system. (She resigned last month.)

Interior Secretary Ken Salazar has elevated the Office of the National Landscape Conservation System to the level of a directorate within BLM, with preservation its priority.

Given that newly exalted status, questions arise. First, who led the campaign for Bingaman's bill?

> That's easy: the Albuquerque-based New Mexico Wilderness Alliance, a purist group out to "re-wild" and restore the Americas to pre-Columbus conditions.
>
> NMWA has been more effective than its $1 million revenue suggests. Its IRS Form 990 shows why: The "affiliations" disclosure shows NMWA "works together on ongoing basis" with three rich Big Green groups: the Wilderness Society (which led the campaign to create the conservation system), the Pew-supported Campaign for America's Wilderness, and, most importantly, the Sierra Club, which can spend limitless sums lobbying legislators.
>
> But the New Mexico Wilderness Alliance dances on the strings of its foundation funders. The top donors are the Wyss Foundation, which also bankrolled Wendy Van Asselt's conservation system campaign for the Wilderness Society, and the Wilburforce Foundation, which specifically donated to "the Organ Mountains Wilderness Campaign."
>
> Top donor Wilburforce ($750,000), based in Seattle, is the money of Gordon Letwin, one of the 12 original employees of Microsoft, which was born in Albuquerque and moved to the Pacific Northwest in 1978.
>
> Letwin quit Microsoft in 1993 to "kick back" with his $20 million fortune. His wife Rosanna now runs Wilburforce with the income from 300,000 shares of Microsoft stock as a money spigot for green groups.
>
> The first of this month, Wilburforce hired Van Asselt as a program officer, giving her a credential on all three sides of an Iron Triangle—activist, government, donor—and illustrating the revolving doors of Big Green.

Okay. So government policy toward energy development and the environmentalists' goals seem parallel, but does that really mean anything? Couldn't this just be coincidence, or maybe just a reflection of the times?

Clearly, environmental groups believe oil and gas are the villains. The goal is to eliminate their usage. If these groups continue to have their way, the car in your garage would no longer be usable. With that wave of the magic wand that grants the environmentalists' wishes, gasoline supplies would run out quickly—like the impact we've felt when a natural disaster shuts down a refinery (only on a much larger scale).

Without the ability to easily fill up the tank, we would be restricted to public transportation—which might have access to the limited fuel supplies. This would work for those who live in town and go to conventional places at normal times. But for those who have an unconventional schedule or who live in rural areas—those places with "rural character, and our unique quality of life," they'd be stuck, unable to get to work, shopping or recreation.

This would also eliminate any vacation travel in which one might engage. A family could no longer hitch up the pop-up and travel a few hours away for a weekend in the woods. And, of course, airplanes would have to be sent to museums. Like the dinosaur, they'd be relics from another era as no airplane can fly on wind, solar, or biofuel.

Some would cheer this point, but let's continue on. At first glance, you might think, "No worry. We'll have to convert to electric cars. This move is long overdue anyway." Electric cars have finally become a reality—though implementation is slow.

This, too, is problematic as 19 percent of America's electricity is generated through the use of natural gas and 3 percent comes from oil[11] —which, with the wave of this magic wand, we would no longer have in abundance. Electric utility companies are worried about the increased need for electricity if electric cars are widely adopted. If a large percentage of the population adds plugging in their car to turning on the TV, powering up the stove, and the heating or cooling that comes with arriving home in the evening, the peak use will be transferred to later in the day and to residential transformers not equipped to handle the use. Major blackouts could occur while the grid is reconfigured—a huge cost that will be absorbed by the ratepayer and taxpayer (you).

Another 50 percent of our electricity comes from coal.[12] While coal is a cost-effective and abundant energy source, for the environmentalists, it has big problems. First, it is one of the largest producers of CO_2. So any group embracing global warming concerns is automatically opposed to coal-fueled power plants. It has been decades since new coal-fueled power plants have been widely built in America. China now has coal-fueled power plants that are cleaner than anything we have in America. Generally, environmental groups are aiming to prevent any new coal-fired plants from being built, while others rejoice in shutting down existing plants. We have yet to develop a cost-effective way to sequester the CO_2 in the emissions. So, with the wave of this magic wand, all those plants would be shut down. Even without the magic wand, the EPA is busy forcing closures.

Environmental groups also oppose coal liquefaction,[13] which means that even though America's coal reserves rival Saudi Arabia's oil reserves, we cannot count on coal for liquid fuels either.

While we will address this more completely in the next section, if we were able to continue, for example, the generation of electricity through wind or solar, the amount of electricity available from those limited resources would need to be reserved for vital services and, maybe, public transportation. Certainly the post-magic-wand limited supply of electricity could not be used to fuel an electric car for going on rounds of errands, hauling kids from one sport or music practice to another, or taking a trip to the mall.

If we are going to have to live without liquid fuels and electricity, what are we going to do for transportation? Remember, one of the groups wanted that "rural character?" You'd get it. Like days gone by, maybe we could all get around on horses! That would surely give us a "unique quality of life."

From the beginning of cities (which were needed to allow people to be close to everything), until the automobile became the primary mode of transportation in 1920s, horses were the conventional mode of getting around. That sounds environmentally friendly and earthy, until you look further. At the turn of the century, the piles of horse manure were a huge pollutant in America's cities. Not only was the stench foul, but the piles shoveled onto every street corner attracted flies and rodents, and, ultimately, disease. San Francisco has a rule on the books that bans piling horse manure more than six feet high at street corners.[14] The resulting filth in the streets produced the custom of the gentleman walking on the "street-side" of the sidewalk when escorting a lady.

Not only is the byproduct of the use of draft animals for transportation a problem; their use created additional waste. A big city had to clear 10,000 to 15,000 horse carcasses from the streets every year[15]—and this is with only about 100 million[16] people in America. Imagine how much worse this would be with the approximate current population of 300 million people!

Like today's vehicles, the horses also need fuel. While they do not run on gasoline or electricity, they do need hay and oats. In 1920, when the horse was the way to travel, one quarter of American farmland was planted with fuel for the horse-based fleet. One horse requires nearly five acres of land a year for its personal fuel supply[17]—as much as is needed for eight people.[18] With approximately 300 million[19] people in America, even if we had only one horse per four-person household, we'd need 75 million horses and 357 million acres of farmland—more than twice the size of Texas.[20] Most of our farmland would be consumed by feeding draft animals. Assuming other countries had not succumbed to the environmental lure, we'd need to import food from them—if we could get it from farm to table before it rotted.

Exacerbating the problem in America, very few households have only one car and many households have only one or two people. The total number of horses needed to maintain anything close to our current lifestyle would be at least twice the one-horse-per-four-person-household number. Horse and buggy are hardly a suitable solution.

Clearly, waving a magic wand and giving environmental extremists all their wishes will greatly alter the way we get around in America. One of the basics of the American way of life is the freedom to come and go as we please, and our cars are a central part of that right. Take away our cars, and you take away our freedom.

When surveyed, 91 percent of Americans responded that they view their cars as a necessity, not a luxury.[21] The car is a symbol of American personal freedom and one the vast majority of us would not want to live without. What else do we consider a necessity that we might have to live without if the environmentalists could wave a magic wand?

Chapter 3

Modern Conveniences: Lifestyle in Environmental Utopia

Are you beginning to get the idea that waving a magic wand to give the environmental groups everything they want would not create the perfect world one might imagine at first glance? The lack of oil and gas would eliminate transportation as we know it. Add to that, the opposition to coal and nuclear and we now have virtually no electricity.

Of course, with a greatly reduced capacity for electricity and transportation, our lifestyle, as we know it, ceases to exist, but let's look at what else we count on today that would be eliminated with the abundant and available energy sources virtually removed from our everyday life.

The Pew Research Center did a survey in 1996 and then again 2006.[22] The Pew survey asked the "Luxury or Necessity?" about different consumer products "designed to help make everyday life more productive, more convenient, more comfortable, more efficient or more entertaining." The results determined that most Americans believe that the following items are a necessity, not a luxury:

- Car
- Clothes washer
- Clothes dryer
- Home air conditioning
- Microwave
- Television
- Car air conditioning
- Home computer
- Cell phone[23]

Each of these "necessary" items has one thing in common: energy is required to both manufacture and use them! They each require either a liquid fuel such as gasoline, or electricity—which is primarily generated from either oil, natural gas, or coal. Since, as we determined in the Transportation chapter, none of these fuel sources is environmentally acceptable, we'll need to use "alternatives" to maintain our accessories that are "necessities."

Some of the most common—and environmentally acceptable—alternatives include hydropower, wind, and solar. Currently hydropower accounts for approximately 6.5 percent[24] of America's energy supply, wind 0.4 percent[25] and solar 0.1 percent.[26] Ideally these numbers will increase, but even the American Wind Energy Association only hopes that wind will increase up to 20 percent[27] of all electricity supply—an optimistic goal, considering that recent figures show it supplying less than 1 percent. They do not expect wind to ever be a "baseload" energy source—which means that it will never be the sole source for electricity and that it needs a primary supply to the energy grid to which wind can be a supplement. Wind will be used as a part of a bigger energy portfolio but it can never be a true "alternative."

To continue life with any of the qualities that we currently enjoy, we will need electricity—lots of it. What could we do to make up the energy gap between the "need" and the above alternatives? Many might think nuclear. It is getting more and more positive nods, as it is the cleanest, least CO_2-producing way to generate the large amounts of electricity needed in America. However, in waving the magic wand, nuclear is out. None of the twenty-five environmental groups studied are "for" nuclear power and several are adamantly against it (this was before the nuclear power crisis in Japan)—with the Nuclear Information and Resource Service's entire goal to create a nuclear-free planet.[28]

But let's assume for a moment that hydro, wind, and solar can create enough electricity for America to continue to have electric power. (This is truly wishful thinking, as some environmental groups are making it difficult to generate electricity by trying to block or shut down wind and hydro.[29]) What would America look like, if it were exclusively fueled by this alternative trio? To meet Greenpeace's goals for wind and solar generation of electricity, we'd have to be installing wind turbines at the rate of more than twenty per hour (and we are already dealing with an increase in the price of steel and a wind turbine shortage[30]) and have photovoltaic cells covering an area as big as the state of Virginia.[31]

There are other, less obvious, problems once we have removed hydrocarbons from life. Every one of the above mentioned "necessities" requires petroleum—and not for electricity, as you might think. They all contain plastics. The casing on your computer and many of the inner workings are made from plastics. So is your cell phone. While your washer and dryer may be metal on the outside, many other parts are made from plastics.

A clever advertisement for the chemical industry, showed many of these "necessities" disappearing from life if we did not have plastics. The ad concluded that we need "chemistry." This is true. But it does not take the next step to tell you that these plastics come from a feedstock (source) of hydrocarbons—coal, oil or natural gas.[32] New research is being done to generate plastics from plant-based feedstock. Design News Magazine declared: "Corn is the new feedstock of choice for plastics aimed at engineering applications."[33] So, now not only are ethanol and food products competing for the ever-more-expensive corn, now the crop is going to have to stretch to provide feedstock for plastics?

As a country, America is already strapped for plastics. The iconic "pink flamingo" died just short of its 50th birthday, as a result of higher prices for plastic resins.[34] Other manufacturing is going overseas where the feedstock costs are lower.

So, without petroleum, many of the things we take for granted—even consider a necessity—will be gone. Here is just a partial list of the modern conveniences most people (72 percent[35]) do not realize are made from petrochemicals:

- baby strollers
- balloons
- cameras
- clothing
- crayons
- credit cards
- DVDs and CDs
- foul-weather gear
- furniture
- luggage
- pantyhose
- toys
- umbrellas

We might, in jest, think we'd be happy without some of these items, but in reality, we count on them.

Remember earlier we decided to pretend that we could still have electricity—from hydro, wind, and solar. Let's also pretend that we can have plastics from corn. We'd still have all of these items we really want, right? Wrong. There is more involved. We may have electricity available, but we have to get it to our homes and offices, which we are also still pretending we could actually reach (probably on foot).

To get the electricity from the power plant to our homes and businesses requires transmission towers and wires to transport it. Copper is the wire on which electricity travels most efficiently. Do we still have copper after the magic wand is waved? Nope. Groups such as the Gifford Pinchot Task Force and Save the Scenic Santa Ritas are aiming to shut down current or prevent future mining of copper and other minerals necessary for energy production.[36] As a result, even though there could theoretically be electricity without petroleum and even though there could theoretically be plastics without petrochemicals, we'd not be able to access the electricity without metals.

Our post-magic-wand existence is getting bleak, isn't it? While there may still be some public transportation, by now most of us are probably using horses for personal transportation—which is using up huge amounts of land for their fuel—and their waste matter is polluting the cities. Without plastics or abundant electricity, we do not have most of the things we consider to be necessities.

At least we have our health. They say without your health nothing else matters.

Transmissions lines: the missing link in the evolution of energy (11 May 2010)

In early 2010, accidents and subsequent deaths resulting from the extraction of traditional energy generated buzz in support of renewables as the obvious energy evolution.

Despite the power density[37] shortcomings of wind and solar (nuclear power: 56W Per Square Meter, solar: 6.7W PSM, wind: 1.2W PSM) and the storage problems associated with delivering the renewable power to the grid, wind and solar are pushed as *the* solutions—propped up by government subsidies. T. Boone Pickens concedes[38] he would not have an interest in wind power without the federal backing.

Though renewables get all the good press, they have been unable to make the leap to mainstream.

Energy has evolved[39] from wood, to whale oil, to coal, to petroleum, to nuclear. There are many who believe that renewables, particularly wind and solar, are the next step in the evolution of energy. But there is a missing link: transmission lines.

As illustrated above, wind and solar are land intensive. To be able to replace an average coal-fueled power plant with solar will require massive amounts of land—which predicates that the installation be "out," away from the people. To get the power "in" to the people will require miles and miles of high-voltage power lines. Somehow, no one seems to think about how the energy is going to get from there to here. Perhaps they think it could actually be transmitted via wi-fi.

There are five obstacles that need to be overcome before renewables can provide America with widespread, utility-scale power.

1) Cost

This is the first hurdle and one that consumers seldom think of, yet the up to two million dollar cost per mile must be calculated into the cost of any new power plant. Acquiring rights-of-way for the lines can increase the cost considerably.

2) NIMBY

No one wants to live near an industrial-looking, coal-fueled power plant. But they have the same Not-In-My-Back-Yard attitude about renewables. The ten-year process for Massachusetts' Cape Wind Project[40] has finally achieved federal approval, though opponents pledge a continued court battle. Additionally, no one wants transmission lines nearby making the new acronym BANANA: Build-Absolutely-Nothing-Anywhere-Near-Anyone. Transmission line installation often requires the use of "eminent domain"[41] as disputes over "fair-market value" arise over the impact of transmission lines on the remainder of the owners' property value.

3) Environment

BANANA leads to the overarching environmental attitude against disturbing the earth. Building the miles and miles of transmission lines, not to mention the land tied up with wind or solar, will require digging, lots of it. In Southern

California, the Sunrise Powerlink project that would bring solar power to San Diego has been trying to build transmission lines for four years. Their biggest opponent has been environmental groups[42] who do not want the lines to go through federal lands.

4) Political Will

Each state's government must decide which they want: renewable energy or the blessing of the environmental movement—which do not seem like they should be mutually exclusive. In New Mexico, Governor Richardson encouraged biomass and then hindered it with the environmental permitting process. In Minnesota, there are myriad regulations restricting power lines, the purchase of power from older power plants, and even those that cross state lines.

5) Cancer Scare

Every couple of years a new study comes out that says that living near transmission lines causes cancer. In 2007, ABC[43] reported on a study indicating an increased risk of blood cancers in people with prolonged exposure to high-voltage power lines. In 2005, CNN[44] covered a British study: children who live close to high-voltage overhead power lines have an increased risk of leukemia.

Each of these five difficulties represents different groups. Getting them all to agree on a single powerline at a single time is an insurmountable task that is stalling renewable installations.

The American Wind Energy Association reports[45] that 2010 had the lowest first-quarter installed generation since 2007. Wind power companies want to build in New Mexico, but are held back by the lack of transmission lines. Maine[46] is ripe for wind power and has commercial wind farms. However, they risk losing a lot of money—and, therefore, jobs—if the state can't figure out how to actually transport the power from the turbines to the grid to the user.

Energy is a tough business. Before we curtail existing energy extraction, we'd better realize, as Obama said on March 31, 2010, "We are going to need to harness traditional sources of fuel, even as we ramp up production of new sources of renewable, homegrown energy."[47]

Chapter 4

Health: The Good Life in Environmental Utopia

Native American Lore offers this formula: "Nobody can be in good health if he does not have *all the time* fresh air, sunshine and good water."[48] If only "fresh air, sunshine and good water" was all that was required for good health, the Environmental Utopia subdivision would be the perfect place we expected. Not that these qualities of life are not essential, but in our complicated world, much more is involved in health. Of course, this statement was made before the industrial revolution when the average life expectancy was about thirty-five years—and fresh air, sunshine, and good water were about all one had for medicine.

Now, the average life expectancy is double what it was before the industrial revolution, and medicine means so much more than the bottle of alcohol-filled elixir the traveling salesman sold in the Old West. Without abundant energy, the field of medicine would revert back to the days of the independent physician. The only way to contact the doctor might be to notify him of a need by a rider on horseback (remember Environmental Utopia has no cars), and the doctor then riding to the location as quickly as possible—which was often too late. His tools were manual and the medicines primitive.

Modern medicine is highly dependent on energy. X-rays, MRIs, and radiation treatments could not be possible without electricity. Even something as simple as a blood-pressure or temperature check, is typically done today with electronic equipment. Not to mention the lights in the hospital that allow emergency surgery to be done at any time of the day or night and the heating and cooling systems that aid in keeping the patient comfortable. Without energy, none of this would be possible. But take this a step further and there are no ambulances to take the injured to the hospital, and certainly no life-flights. Helicopters are not a part of Environmental Utopia.

Political pressure to fix the energy problems of the future (10 February 2011)

The coldest temperatures[49] in the contiguous states created a "state of emergency,"[50] due to a disruption of natural gas delivery. Power plants were shut down, workers sent home, schools closed, and shelters set up for those without heat. This was Thursday, February 3. Many people remained without heat a week later, while temperatures reached as low as thirty-degrees-below zero.

This was in New Mexico but other states in the Southwest faced rolling blackouts and a variety of energy-related emergencies. This same problem could face the nation, if we continue down the path we've been pursuing for our energy supplies.

Businesses, residents, and hospitals received word[51] that there might be natural-gas disruptions throughout the day. This did not mean that there was, as many believed, a shortage, but rather rolling blackouts in Texas were causing pressure problems in the pipelines—the delivery system. There may have been enough gas to fire up your cooktop, but not enough pressure for a furnace. The cold weather and lack of heat combined to create a state of emergency.

It turns out that the local utility company, which gets much of its natural gas from New Mexico, also gets some gas from Texas to spread out its risk. That natural gas comes through a pipeline that needs pressure to keep it moving. Due to Texas' problems, the compression stations along the way had no electricity to pressurize the natural gas, hence the "disruptions."

Why did Texas not have enough electricity? Investigations[52] will determine the true cause of the energy emergency and myriad explanations have been offered. But here is what we do know:

- Cold weather put a strain on the system leading to failures that threatened peoples' lives and disrupted everything.
- Texas' heavy reliance[53] on wind power[54] may have been part of the cause.
- There were no problems[55] with nuclear power plants in the area.

We do know the same thing could happen almost anywhere and almost anytime. It comes down to a lack of reliable electricity.

How could America, the wealthiest and most technologically advanced nation in the world, be faced with life-threatening energy shortages? Bottom line: our policy toward power generation—a policy that must be changed.

First, funding and political favor have supported the intermittent and expensive electricity sources of wind and solar. Neither can be turned on when needed. The wind must blow and the sun must shine. On frigid nights when the temperature drops dramatically and people turn on electric heaters, the wind is still and the sun doesn't shine—making them virtually worthless, when the energy is needed most.

Next, the building of new, high-tech coal-fueled power plants has been impeded or totally blocked. In America, we are limping by with a generation of coal plants far behind those currently being built in China. The regulations on anyone who tries to build a coal plant, as President Obama promised, "will bankrupt them"[56]—resulting in very few new coal-fueled power plants. The older plants had weather-related problems and had to be shut down. Their closure put more demand on the gas-fueled plants—which also strained the natural-gas delivery to commercial and residential customers.

The only power source that had no problems was nuclear. Despite Obama's early pledge[57] to encourage the building of nuclear plants through loan guarantees, nothing has moved forward on that front. Regulations make their construction nearly impossible and add significantly to the cost. The political atmosphere is a barrier to private funding. It has been nearly twenty years[58] since we've had a new nuclear power plant come online.

Texas has the highest wind capacity[59] (installed or under construction) in America. They also have what is called a Renewable Portfolio Standard (RPS), which requires the state to have a set amount of renewable energy by specific dates. Twenty-nine states[60] have an RPS or a non-binding goal for adoption. If renewable proponents have their way, Senator Bingaman's plan for a national Renewable Electricity Standard[61] will become law and every state will be investing in expensive energy that is not available when really needed and blocking the electricity that can truly power America.

The problem last winter in New Mexico was lack of pressure in the pipelines. To fix energy problems for the future will take political pressure.

If a person had a heart attack, for example, and was able to get to the hospital via horse and buggy, he or she would not be able to receive the life-giving capabilities of a pacemaker or an artificial heart. Each of these modern inventions would not be possible without plastic—and plastic is made primarily from hydrocarbons.[62]

Without plastic, the changes that would take place in the fields of health and medicine are myriad, as it plays a much greater role than we think—if we *do* think about it. In the hospital, plastic's presence ranges from sterilization trays and advanced housing materials, to endoscopic surgical instruments and blood filters—and nearly everything in between. Most products synonymous with emergency medicine could not be produced without plastic: IV bags, syringes, and portable, lightweight equipment, such as cardiac monitors, are just a handful of examples. Prosthetic devices have revolutionized the medical field; plastics have given everyday people in need of prosthetics increased mobility and artificial limbs with an astonishingly lifelike appearance.[63]

Disposable plastic products have dramatically reduced infection rates and the spread of disease. Plastic packaging and containers help keep medicine and instruments sterile and secure at your doctor's office. At home, plastic child-proof caps prevent accidental deaths, and our plastic eyeglasses help improve our vision. Every day, modern medicine depends on plastic to help keep us healthy and lower costs.

Then there is the equipment that adds to our health and comfort—much of which uses both electricity and hydrocarbon-based plastics. Look at the treadmill or stair-stepper you might use at the gym or at home. They require electricity to work, and they have plastics for comfort. The gym shoes and clothes you wear, while using the equipment at the health club, are probably all petroleum-based. At the doctor's office, think about what is under that paper they pull over the exam table—a vinyl-covered, foam cushion. Plastic is at work in our health and comfort in ways we'd overlook unless they were absent—as they'd be in Environmental Utopia.

While medicine is important, plastics play a big role in our overall health and life. From the manufacturer to the grocer and to our dinner tables, airtight plastic packaging helps keep foods fresh and free from contamination.[64] Without plastics, our food supply and its safety would be jeopardized.

Without modern transportation, our food selection would change drastically, and this would impact our health. Some environmental groups suggest that we not transport food and that people buy from local farmers markets only.[65] We would not have fresh fruits and vegetables at all times of the year—and certainly not the vast selection we find in the local grocery store in refrigerated cases.

Give thanks for energy (23 November 2009)

I boarded a flight heading to a speaking engagement later in the day. The flight attendant commented on my attire. I told her that I was a "speaker" and therefore could dress with a bit more bling than the average person. "Oh, what do you speak on?" "Energy," I replied. "Great, I used to be a nutritionist," she responded. I told her that it wasn't that kind of energy.

As we head into Thanksgiving, you may think you need lots of her type of energy—and you'd be right. But without my kind of energy, you'd need a whole lot more of her kind of energy to create the "old-fashioned" Thanksgiving that so many of us picture when we think of the nationwide holiday.

One of the big traits of Thanksgiving is the entire multi-generational family gathered around the table. Back in the day of the picture-perfect holiday, travel meant hitching up the horse and wagon. Today, to accomplish this, family members often have to travel great distances to get to the site of the big meal. Thanksgiving is reported as the busiest travel season—whether by auto or air. But even before the travel takes place, energy is a big part of the picture.

The travel has to be planned. Air travel takes a visit to one's favorite travel website. Travel by land often requires a MapQuest® search for the best route. Both need energy to function. Then when the actual travel takes place, regardless of the method or distance, fuel is needed to make the trip possible.

Even the big meal takes more energy than one might assume. First, the turkey has to be raised (I'll not belabor each phase of energy used there). Then to get it to the store in a safe and sanitary manner, requires refrigeration and transportation—both are energy dependent. Once at the store refrigeration, is, again, important. To go to the store to make your selection demands fuel.

Let's jump to the big day. Most people stuff the bird and cook it in the oven—although the fried turkey has increased in popularity. Either way, energy is required for cooking—natural gas, electricity or propane. And, that does not include the veggies, the mashed potatoes (that need a mixer), or the freshly baked rolls. The Thanksgiving feast typically includes some sort of salad. At my great aunt's home in Massachusetts, salad was green Jello with chopped celery and a dollop of mayonnaise. In modern homes, the salad is usually lettuce-based. Either way, energy is needed to keep things fresh and cool.

Once the meal is ready, many people use an electric knife to cut the turkey and a hot plate to keep things warm while the final preparations are made. Both need energy.

Around the table, the ambiance may be created the "old-fashioned" way with candles and a flickering wood-fueled fire. But even fire is energy—the first used in civilization. But maybe you have music playing on the stereo—tunes downloaded from iTunes (thanks to energy).

Post-meal, fat and happy, many households will retire to the sofa to watch the big game of the day. Once again, energy is a central feature. It gets all the on-site participants to the stadium. Energy lights the stadium and powers the television cameras. Perhaps the image gets to your home via satellite or cable. Neither is possible without energy. Once in your home, that flat-screen TV needs electricity and rare earth to give you that great picture. The beer you're drinking is nice and cold, thanks to energy.

But it is not over yet. In most homes, while the men watch the game, the women clean up. Whew! The dishwasher makes it so much easier. And the hot water coming straight from the tap is expected. Once again, energy.

If you are the cook, by the end of the day, you are ready for all of those people to head home. You are lacking energy and are ready to crawl under your electric blanket—all warm and snuggly.

Once you start thinking about it, you can see myriad other ways that energy makes your Thanksgiving the picture-perfect event of which you've dreamed. Maybe you used your digital camera, your computer to view the shots, and your printer to print out copies for everyone. You get the picture.

When you come to the table and bow your head to give thanks for the family, friends and food—don't forget the energy that made it all possible.

Without electricity to run the refrigerator we have at home, we'd only be able to eat today what we harvested from the garden—or shot or captured—or what we canned during the last harvest. Not to mention how much harder and less productive the growing and harvesting process would be without the modern, gas-powered machines.

Without energy, we would surely have that longed for "rural character," but is that really how we want to live? Even those who live in rural areas have most of the above-mentioned energy-using appliances, vehicles, and amenities.

Chapter 5

Housing: Open Living in Environmental Utopia

As we have systematically peeled away many of the elements of life today, what do we have left? Maybe we would not have electricity nor all the benefits it gives us every day (including over 40 percent of a building's electricity for heating or cooling, and a little over 10 percent lighting); maybe we'd not travel far from home because long trips via horse and buggy are tiresome; and maybe our life expectancy would be shorter without the benefits of modern medicine, but we'd at least have our house to keep us protected from the elements and safe from wildlife attacks, right? Negative.

When I was first developing this idea, a contrarian snapped at me and said, "Well, I have a house now. This wouldn't be a problem." And you may still have a house in Environmental Utopia. But in our greatly reduced energy world, you couldn't mow your lawn with your powered mower—in fact, you wouldn't even be able to water it. Your current home would quickly begin to look shabby and run down. Without electricity available for water processing and pumping, you would soon need to leave your home to be closer to a water source—which eliminates the idea of living in most cities to be close to everything. But then, everyone else is going to have to move to the water source as well, so you will not be lonesome.

The Environmental Utopia subdivision will have homes that are unique. You'd assume they'd be "green" and "sustainable." But you'd be wrong.

Let's start with the obvious: wood. Ninety-four percent of houses in the U.S. are built of structural lumber.[66] But in this new world, wood would not be available as numerous groups oppose logging and are currently working to stop it here or there.[67] Many groups—including Wild Earth Guardians and Western Watersheds Projects—are opposed to harvesting trees of any size, therefore making sustainable forest management difficult, if not impossible to accomplish. This creates insect and disease problems. Forests choked with too many trees lead to fires like the Cerro Grande, Lake Tahoe fires, and Southern California's late 2007 fires or the more recent 2011 Wallow Fire, which burned more than half a million acres—which then create a different set of problems. Under a limited fuel-supply scenario, catastrophic fires would be the norm, and they would be larger and more intense. They'd burn for days on end as there would be no mechanical equipment to suppress them.[68] These catastrophic fires are a major producer of greenhouse gases with an estimated 4.8 metric tons of CO_2 for every acre of coniferous forest consumed.[69]

In some parts of the country, homes are frequently made from cinder blocks. Besides the fact that energy is required to make cinder blocks, we would not have them either, as they are made from cement—and cement (made from limestone) and its components are under attack by groups such as Earth Justice.[70] This would even mean that a home built from rocks could not be held together with cement.

With our current housing left behind in suburbia, what is left to live in? To quote Chief Flying Hawk, "The tepee is much better to live in: always clean, warm in winter, cool in summer, easy to move..."[71] While the tepee was an efficient mode of housing for the nomadic life of the plains Indians, I can't imagine Al Gore fitting his 10,000 square-feet of stuff into a tepee, can you?[72] I don't want to live in a tepee, and I am pretty sure neither you nor Al Gore wants to go tenting on a permanent basis. They'd be leaky in the rain. And I think that they would not really be, "always clean, warm in winter, cool in summer."

So we find ourselves down to caves. In America we'd need 75 million caves with four to a household. I doubt the caves would be "always clean." They'd have dirt floors. Carpet today, like much of our clothing, is made from petroleum.[73] Speaking of clothes, where would they come from? You can't run down to Walmart, Chico's, or J. Crew to pick up your favorites. Unlike the GEICO® cavemen who live in our modern world, we would not have the convenience of ready-made fabric, let alone clothes.

Then, remember we mentioned that the caves would need to be near a fresh water source. Without electricity for pumping, processing and purifying and without plastics or copper for piping, we would not have the ease of clean water from simply turning on the faucet. But we need water for health and cleaning. So we'd have to be near a stream, pond or lake. However, lakes, rivers, springs, ponds, or streams can be contaminated with feces from animals, resulting in giardiasis—a diarrheal illness caused by a one-celled, microscopic parasite.[74] The U.S. Government recommends that backpackers, hikers, and campers not drink unfiltered or untreated water—water that has not been boiled, filtered, or disinfected with chemicals.

But let's be optimistic. Remember, we are dealing with an element of magic here. Suppose you find a cave near a fresh water source that is not contaminated with giardia. Maybe this cave is in an Eden-like setting, and the area is teeming with fresh food sources and abundant trees to provide for firewood—over which you can cook your food and create a toasty fire to keep you warm in the winter.

Whoa! You cannot cut down those trees, remember? OK, since we are talking "utopia," let's assume this cave is in a warm locale so you do not need fire for heat. (Since you'd not have access to fire trucks or borate bombers to fight fires, it is a good thing you'd not need fire.) You do have the true essentials: food—albeit raw—and water. Like the wildlife leaving feces in the water, you will need to do something with your personal waste matter. If you were simply camping in the mountains for the weekend, you could find a convenient log and dig a hole, which you'd cover up before leaving the site. However, remember with the wave of our magic wand, we gave the environmental groups what they want—and this is now your life. With 300 million people in America looking for caves near a fresh water source, the chances of your being alone in those woods is slim at best. There will probably be more people in the area producing waste matter than the ground can absorb. Without plumbing and electricity, there is no sewage processing. Waste matter will not be converted into "non-potable" water that can be used to water the plants along the freeway. (Of course, we will not have freeways.)

If you think about it, you'd realize (like the camping example in the introduction) you will have no time for a job. You will be spending all your time trying just to survive. Life as we know it would disappear. A giant worldwide recession would occur. But then, without computers and the Internet, the whole "global" community would cease to exist.

Yuck! It's rapidly looking like the environmental groups have fed us a grocery cart full of misinformation and propaganda. The world would not be better and brighter, if they were in charge. Mountaingate, the community built around a golf course, built on top of a landfill, is looking really attractive.

Chapter 6

Only A Few Models Left

Yes, the Environmental Utopia subdivision here may seem extreme, but we have barely touched on the full impact of what would happen if the "magic wand" really could be waved. There are many other elements we could have developed and documented in this narrative, but the end product would be so long that we fear no one would want to read it.

Here's a quick idea of some of the additional areas we could expand upon.

- Regarding wind power, this utopia would have a problem in that mining the ore to make the turbines (and the actual manufacture of them) requires energy.
- If corn is the "feedstock of choice" as is mentioned, we have not delved into how much agricultural and wildlife habitat acreage will be needed to produce that corn. Nor how much water, fuel, insecticides and fertilizer we will need to grow and transport, not to mention the fuel needed to convert the sugars to fuel.

- Speaking of corn, we have barely mentioned ethanol. This topic is so large that it requires analysis of its own—which is provided in Section 3, Chapter 12.
- Since we know people would revolt before we really got to the cave stage—though some ecologists prefer we reduce the population by dying off,[75] the initial changes would hurt the poor, the elderly, and minority families the most.

While we all want a green earth—clean air, fresh water and a safe food supply—stopping, opposing, denying, and blocking are not the ways to get it. Huge strides have been made since the dawn of the industrial revolution. Despite increases in the typical activities that produce pollution, America is much cleaner now than it was twenty years ago. Since 1970, our population has increased 42 percent and our energy consumption has gone up 48 percent. The number of miles we drive each year is up 178 percent. At the same time, our American ingenuity has been at work generating a 195 percent increase in the GNP. With these figures, it would be only logical that pollution has increased. However, the truth is that, despite this growth, our aggregate emissions are approximately half of what they were![76] Our air is clearer. Our streams are much cleaner—while they may have giardia, they are teeming with fish.[77]

> The U.S. is certainly one of the cleanest, more environmentally responsible nations in the world. Virtually, no European country can boast cleaner waters, or more pristine rural landscapes or air quality.
>
> Michael Economides, Author, ***Energy and Climate Wars***

Yes, there do need to be regulations, as every population has its "bad actors." Yes, we do need to be aware of the damage certain activities create. We also need compromise to protect good people from irresponsible choices. Alternatives that sound better may not be the best option. Extreme environmentalism has just as many downsides as the progress-at-any-cost-extreme. No one wants drilling in his or her neighborhood, but we all want inexpensive gasoline. We all want to turn on the lights at will, use our computers and our iPods, be cool and comfortable in the summer and toasty warm in the winter, but no one wants a power plant in the state—and the Kennedy family certainly doesn't want wind turbines messing up their high-priced ocean view or dropping their property values. Environmental groups are even fighting the installation of wind farms, charging that greedy companies are "seeking sites with the most wind..."[78] In New Mexico, Santa Fe residents put a moratorium on oil and gas exploration,[79] despite the fact that these resources provide approximately 50 percent of the state's revenues—and we are one of the poorest states. It is like the old saying—there is no "free lunch." We cannot have the energy we want without some elements we do not want.

When someone in front of the grocery store asks you to sign a petition to stop plans for a new power plant, or a new refinery, or to stop logging here or mining there, think of the bigger implications. While our world is not perfect, I think you'd rather live here—wherever your "here" is—than in "Environmental Utopia."

Let's get real! Let's look at the energy reality.

Do you prefer the "necessities" of life? Do you like your car, your microwave, your health, and your house? Now that you are aware of the changes that the seemingly innocent desires for a green earth could force upon your life, it is time to do something about it. If you like the American way of life, don't let the environmental groups take away energy with "nice" sounding salvation plans. Most of them have unattainable goals or goals that do not factor in the complete picture. They fight to stop a perceived evil, without looking at what else will be impacted. Which portions of your lifestyle—or your children's or grandchildren's lifestyle—do you want to give up in the name of saving the world from this or that speculative environmental cataclysm?

Contact your elected officials and let them know you want energy—responsible energy! Ask them to allow more exploration and innovation in America. Tell them you want to tap into energy sources currently or in the process of being made off-limits in America by the Endangered Species Act or plans to lock up resources under the guise of a national monument. Don't let the squeaking wheels of the environmental groups continue to dictate policy. They'll regulate us back to the Stone Age!

We the people, we the consumers, have the power. Call your elected officials; jam the phone systems! We have power over our elected officials, and we can vote with our dollars. Don't buy things that waste energy, but look closely, as things are not always as they appear.

As a consumer, you can demand efficiency. The government doesn't need to mandate it—though they do currently. Consumer demand is what hurt the American auto industry and catapulted the imports to the top sellers. Now Detroit is desperately trying to catch up. An ad for Hyundai said, "It is not that complicated. If gas costs a lot of money, let's build cars that use as little of it as possible." This is called market demand. It is *we, the consumers,* voting with our dollars.

The Great Green Lie

Louis Prudhomme

Today's "green" efforts can be likened to Hans Christian Anderson's classic children's tale *The Emperor's New Clothes*.

You may recall how the mischievous tailors played on the vanity of the Emperor and his court to create a hoax for profit. Sound familiar? Like the Emperor's advisers, the politicos pretend to "see" what they are told is real, while the people aren't buying it.

The players today may be a little different, but the story is essentially the same. In the classic, the weakness is the Emperor's vanity. In the modern version, global warming creates a sense of guilt over potential harm to nature—"weavers" fabricate the idea that we must choose between energy and the planet. This time, more than the Emperor's vanity is at stake and the plot revolves around green energy—with wind and solar as the key to a utopian future. The Great Green Lie is that we need to end our dependence on traditional fuels.

Leaving fiction behind, the truth is that throughout Earth's 4+ billion years, the climate has been both warmer and cooler; the percentage of carbon dioxide and water in the atmosphere has varied considerably. But, like the classic tale, many pretend to see the "truth" the "weavers" put forward, lest they be thought stupid or unfit for their posts.

In the '70s, clean air and water laws were implemented based on local smog and pollution. Today, U.S. companies are examples of ingenuity in their management of emissions. More energy-efficient methods of using resources such as coal, oil, gas, and uranium have been developed. The exhaust or releases are often cleaner than the intake, and certainly cleaner than those in other large industrial countries such as China and India.

The objective of the 1970s legislation was the control of real poisonous releases such as carbon monoxide, sulfur dioxide, lead, etc. Now, the great villains are naturally occurring carbon dioxide (CO_2) and water vapor (H_2O)—known as greenhouse gases (GHG). If all mankind disappeared tomorrow, GHG would remain virtually the same. While solar and geologic phenomena are a greater influence on GHG than the human use of hydrocarbon fuels, the storyline is now about managing individual and national behavior through taxation and rationing. The proponents of such policy have created an artificial crisis based more on guilt than on proven science.

Today, bureaucrats, like the emperor's advisers, defer to these supposed intellectuals to impose higher taxes and more stringent regulations on energy producers while government subsidies pour into building underdeveloped *dependent technologies*—electric cars, solar and wind "farms." (Note these all depend on petroleum or utility-scale electricity to make them possible.)

To know whether or not we are being duped, like the child in the classic tale—"he hasn't got anything on"—we must make honest observations.

- The "renewable" technologies depend on energy and raw materials that must be extracted from the earth.
- It can cost more to produce and operate a Solar Cell or a Wind Turbine Generator than the energy it produces.
- Raising the price of primary energy sources will then raise the cost of the solar and wind technologies.
- Cap and trade doesn't really help anyone; it just moves the problem and raises costs.
- The Chicago Climate Exchange is like ENRON, except the artificial trade commodity is carbon dioxide, a compound necessary to the earth's survival.
- The United States is on track to emulate Spain's failed "Green Economy."

When we stop pretending that we see the "lie," we discover that the American public has been swindled. Solar and wind technologies are far from being able to carry the energy load for large industrial countries. Unfortunately, the "storyline" prevents us from seeing true solutions.

Our energy needs continue to grow and only nuclear power—successfully operated by the U.S. Navy without incident for nearly sixty years—has the ability to truly power the U.S., yet it is shunned here, while adopted in France, Germany and Japan We must start now as it will be decades before we can create the infrastructure to replace coal power plants with nuclear.

Cost-effective, available energy, coupled with ingenuity and industry, has been the power that has made America great.

Now, Americans are expected to produce energy using the same material that the weavers used to dress the Emperor—stuffing their pockets while producing nothing. If the current legislative trend continues, the rest of the world will be laughing, as America will have no clothes.

Have your eyes been opened now that you see the complete energy picture? Then spread the word. Tell your friends and family. Look around. Energy is everywhere. Energy makes America great!

Take away energy, and you take away freedom.

Section Two

Who are the Environmentalists?

There are hidden contradictions in the minds of people who "love nature" while deploring the "artificialities" with which "man has spoiled 'Nature.'" The obvious contradiction lies in their choice of words, which imply that Man and his artifacts are *not* part of "Nature"—but beavers and their dams *are*. But the contradictions go deeper than this prima-facie absurdity. In declaring his love for a beaver dam (erected by beavers for beavers' purposes) and his hatred for dams erected by men (for the purposes of men) the "Naturist" reveals his hatred for his own race—i.e., his own self-hatred.

In the case of "Naturists" such self-hatred is understandable; they are a sorry lot. But hatred is too strong an emotion to feel toward them; pity and contempt are the most they rate.

As for me, willy-nilly I am a man, not a beaver, and H. sapiens is the only race I have or can have. Fortunately for me, I *like* being a part of a race made up of men and women—it strikes me as a fine arrangement and perfectly "natural."

Believe it or not, there were "Naturists" who opposed the first flight to the old Earth's Moon as being "unnatural" and a "despoiling of Nature."

From the Notebooks of Lazarus Long" in Robert Heinlein's 1973 novel, ***Time Enough for Love***.

Chapter 7

Smash the Watermelons

So, who are all these evil-appearing "environmentalists?" Should all Birkenstock wearers be suspect? What about the lady at the grocery store with a canvas sack for her purchases?

There is a difference between those of us who care for the earth, want to use it wisely, and believe in recycling—and those who are in decision-making positions, setting policy and/or funding the programs. I believe most Americans, including me, fit into the first group. We learned not to be a "litterbug," and we surely want clean air and water. Personally, I do not like collecting up all those plastic bags from the grocery store. It seems to be just plain wasteful. I carry a variety of fabric bags in my car and try to remember to actually take them into the store. I use most pieces of paper three times: front, back and shredded for packing materials. I live in a rural area on ten wooded acres. I am practically off-the-grid, as both my water and propane are delivered in trucks. I have to remember to call and request delivery service—sadly, sometimes I forget and run out. Does that make me an environmentalist? I think so. But I am not pushing an "environmentalist" agenda. Nor are the majority of Americans who value the earth and all that it has to offer us.

We need another word, one that sets those of us who care for the earth apart from those pushing an extreme agenda. My friend and mentor Ron Arnold uses the term "Big Green" to represent the combination of activist, government, and donor working together to change policy—essentially creating an iron triangle. Big Green is an apt description as the combination does effectively communicate something bigger than the lady with the canvas bag in the checkout line. Like big labor or big business, Big Green carries the connotation of power.

Personally, I like the idea of a "Watermelon." I did not create the moniker; I did not even understand it the first time I heard it. I had to have it explained. "Green on the outside like environmentalism, red on the inside like neo-communism or neo-socialism." Ah! For me, that stuck. I've used the term in speaking and people seem to get it. I have bumper stickers I give out at my speaking engagements that say, "Smash the Watermelons" and pens that are green on the outside that write with red ink. Both are very popular.

Whatever label you like, keep in mind that we are talking about something bigger, something organized, and something with plans greater than saving polar bears or spotted owls. In the next chapter, you will find information regarding some of these Big Green groups, the key players, their leadership, and their funding. What you'll find is something with plans to fundamentally transform the United States of America.

Iron Triangle

A Glossary of Political Economy Terms, Paul M. Johnson

The closed, mutually supportive relationships that often prevail in the United States between the government agencies, the special interest lobbying organizations, and the legislative committees or subcommittees with jurisdiction over a particular functional area of government policy.

As long as they hang together, the members of these small groups of movers and shakers tend to dominate all policy-making in their respective specialized areas of concern, and they tend to present a united front against "outsiders" who attempt to invade their turf and alter established policies that have been worked out by years of private negotiations among the "insiders." The middle-level bureaucrats who run the agencies may use their special friends in Congress to block the efforts of a new President or a new Congressional majority leadership bent on reforming or reducing the size of their agencies. The Congressmen and Senators on the oversight committees can count upon their friends in the agencies to continue "pet" programs and pork-barrel projects important to their local constituencies or even to do special favors for their political supporters and financial backers. Lobbying organizations provide useful information to the committees and the agencies, provide campaign support for the relevant Congressmen, and often help to mobilize public opinion in favor of larger appropriations and expanded programs for "their" part of the government bureaucracy. In return, they tend to be consulted and carefully placated when new laws or administrative regulations or important appointments affecting their special interests are being made. These triangles are said to be "strong as iron" in that these mutually supportive relationships are often so politically powerful that representatives of the more general interests of society are usually effectively prevented from "interfering" with policy-making altogether whenever their concept of the general interest runs counter to the special interests of the entrenched interest groups, bureaucrats and politicians.

Whether we refer to them as Big Green or Watermelons, there are hundreds of these groups[80] influencing policy in America. Here, throughout the pages, we've highlighted only a handful—enough that you know we did not make this up, but, hopefully, not so many as to be redundant and boring.

Top ten environmental groups

Green groups get gobs of greenbacks

Mark Hemingway, *Washington Examiner*, 29 November 2010

Just because an organization is a nonprofit, that doesn't mean it's poor. Environmental organizations are thriving and have considerable financial resources at their disposal, with revenue coming from government grants and contracts, direct-mail and Internet-fundraising campaigns, foundation grants, and gifts from rich activists.

According to the latest tax filings, here's the net worth of some of the nation's largest and most prominent environmental organizations. [Representing just over $8 Trillion.]

1. Nature Conservancy–Total assets: $5,636,393,924
2. The Conservation Fund–Total assets: $451,178,482
3. World Wildlife Fund–Total assets: $426,048,663
4. Trust for Public Land–Total assets: $399,026,229
5. Conservation International Foundation–Total assets: $370,034,224
6. National Audubon Society–Total assets: $337,695,958
7. Natural Resources Defense Council–Total assets: $232,276,696
8. Environmental Defense Fund–Total assets: $145,765,426
9. Sierra Club Foundation–Total assets: $107,928,024
10. National Wildlife Federation–Total assets: $69,448,048

We questioned the motives. As we've already seen, energy is essential to the life we live. Following the actions of these myriad groups, it appears that their plans will end the America we know and love. When you study this day in and day out as I do, you have to wonder, "Why are they doing this?" Some do not seem to like any energy source—many even oppose the so-called "darlings" of the energy world: wind and solar.

"Environmental Organizations, many of them the offshoot of anti-establishment, radical movements of the 1960s, may have progressed to wearing a suit and tie but they still show a stark aversion to any development. There is virtually no commercially viable energy project of which the green movement would approve."

Michael Economides, author, ***Energy and Climate Wars***

I learned how they use the courts, case-by-case, permit-by-permit, to increase the costs of an individual or company using one's own property, with the goal of discouraging the development of that property. There appears to be an aversion to someone benefiting from his or her own personal property—as if it should be used for the collective good.

Permitting is used as a tool to stop development. This can happen because a person who owns a piece of property needs a permit to do anything on it—even building a house. Because the permitting process is so backlogged, the property has a much higher value, if it is already permitted. Due to department staffing shortages, many agencies are months behind in permitting.

There seems to be a simple solution. With all the stimulus spending, use the funds to hire more people in the departments that are backlogged. That would be the solution if you wanted growth. But that is not the goal of most agencies today. The permitting process is complicated for a reason.

During the Clinton administration, many of the people in the Environmental Protection Agency, and other like departments, were personal appointees of Al Gore. They deliberately derailed the regulatory process to put an end to economic development. I have certainly seen the regulatory process deter the development of oil and gas resources in America.

The true environmentalists want people to move into cities where they are easier to control. They want to get rid of inefficient human patterns—which is what happens when people live in rural locations. Think about it. Many of the ideas lauded by environmental groups: electric cars, community gardens, and mass transit, for example, work only in cities.

As I learned more and more about the environmentalists' efforts, it became clear that "control" was really the issue. A complete picture began to emerge and the painting looked a lot like a watermelon: green on the outside and red on the inside.

Of course, the environmentalists do not fit the true Marxist model, but in the general sense, the idea of "red" explains a lot. Columnist George Will says, "Today's 'green left' is the old 'red left' revised." In his op-ed on polar bears,[81] he states: "The left exists to enlarge the state's supervision of life, narrowing individual choices in the name of collective good. Hence the left's hostility to markets. And to automobiles—people going wherever they want whenever they want." This explains the "red."

George Will continues, "The green left understands that the direct route to government control of almost everything is to stigmatize, as a planetary menace, something involved in almost everything—carbon. Environmentalism is, as Lawson (author of ***An Appeal to Reason: a Cool Look at Global Warming,***) writes, an unlimited 'license to intrude.' 'Eco-fundamentalism,' which is 'the quasi-religion of green alarmism,' promises 'global salvationism.'"

This increasing government theme is repeated in the writings of Fred L. Smith, Jr., President of the Competitive Enterprise

Institute. In a 2004 article titled, *Eco-Facism Going Global,*[82] he says:

"We can say this for environmental activists—they are persistent, constantly developing collectivist schemes to increase the size and scope of government. And they are becoming more ambitious.

"Not content with giving more power to Uncle Sam, they now seek to give greater power to the United Nations, a move that would seriously undermine American sovereignty and pave the way for top-down global wealth transfer schemes that would make the Great Society look small. Let's take a look at two such schemes.

"The Law of the Sea Treaty: This treaty would give the U.N. the power to regulate and tax deep-sea mining and redistribute the receipts to Third World governments. It would simultaneously legitimize the principles of green collectivism and global income redistribution. President Reagan rejected the treaty in 1982, but a loose coalition of interest groups and activists are now working to revive this monster.

"The Kyoto Protocol on 'climate change': This treaty seeks to restrict global greenhouse gas emissions, including those of carbon dioxide—the inescapable by-product of energy production. Kyoto's backers ignore the lack of evidence for catastrophic warming and the dire economic consequences of an energy-starved future, but no matter. They seem most interested in Kyoto's wealth transfer potential. Statements by one U.N. official suggest that if Kyoto took effect, the U.N. would ask rich countries to provide 'funding' to poor countries to "help [them] better adapt to climate change" in the name of global "equity."

"Notice how these two eco-collectivist treaties would give the U.N. control over large sums of money. The Iraqi oil-for-

food program scandal has given us a stark lesson on how the U.N. handles funds. As P.J. O'Rourke said about another group of bureaucrats, giving the U.N. more money and power would be "like giving whiskey and car keys to teenage boys."

Are the pieces of the puzzle beginning to fit together? This thinking is totally foreign to my thinking. While I could see what was happening today, I still had to wonder about the root. I'd heard many people mention something like, "many of the environmentalists operate as a result of a 1960s and 1970s mindset they had never grown out of." I graduated from high school in 1976, so I am a product of that generation, but I don't think like that. I needed more to understand the root thinking of these Watermelons.

One of my mentors suggested that I read a specific book written in the early 1970s: ***Cows, Pigs, Wars and Witches***. I wondered how a book by that title could help me on the quest for understanding this mindset. The subtitle is ***The Riddles of Culture***. Still no clue. I was advised to hang in through the whole book as it came together at the end. It was Chapter 11 when I finally got it.

The Watermelons—those who are green on the outside and red on the inside—*are* operating on the remnants of the '60s and '70s. Back in the day, the mindset was, according to ***Cows, Pigs, Wars and Witches***, referred to as the counter-culture or Consciousness III.

One of the first things that caught my attention was this description of counter-culture: "Counter-culture celebrates the supposedly natural life of primitive peoples. Its members wear beads, headbands, body paint, and colorful tattered clothing; they yearn to be a tribe. They seem to believe that tribal peoples are non-materialistic, spontaneous, and reverently in touch with occult sources of enchantment." While much of that comment is totally '70s, the idea of the primitive peoples, of the tribe, seemed to have a current quality to it. It helped to understand the apparent hatred of modern development and the inherent desire to go back to caves.

Now that I was clued in, I went back a few paragraphs—reading with a different focus. I encountered this general introduction of the mindset:

"The unexpected resurgence of attitudes and theories long held to be incompatible with the expansion of Western science and technology is associated with the development of a lifestyle which has been given the name 'counter-culture.' According to Theodore Roszak, one of the movement's adult prophets, counter-culture will save the world from the 'myths of objective consciousness.' It will 'subvert the scientific worldview' and substitute a new culture in which the 'non-intellective capacities' will reign supreme. Charles A. Reich, another minor prophet of recent years, speaks of a millennial state of mind, which he calls Consciousness III. To achieve Consciousness III is 'to be deeply suspicious of logic, rationality, analysis, and of principles.'

"In the lifestyle of the counter culture, feelings, spontaneity, imagination are good; science, logic are bad. Its members boast of fleeing 'objectivity' as if from a place inhabited by plague.

"A central aspect of counter-culture is the belief that consciousness controls history. People are what goes on in their minds; to make them better, all you have to do is give them better ideas. Objective conditions count for little."

Ah. No wonder the fact that you cannot power the world on wind or solar power doesn't faze the Watermelons in their insistent push for ineffective systems. They believe that if they say it enough, it will be so. "Science, logic are bad." "Objective conditions count for little."

See if you don't agree that the following sentences from ***Cows, Pigs, Wars and Witches*** doesn't sound like the Watermelons we have come to know. As you read, try replacing the word "superconsciousness" with the word "Environmentalism." "Its advocates insist that superconsciousness can make the world into a more friendly and more habitable place; they see flight from objectivity as a politically effective way to achieve an equitable distribution of wealth, recycling of resources, abolition of impersonal bureaucracies, and the correction of other dehumanizing aspects of modern technocratic societies. … Capitalism, the corporate state, the age of science, the Protestant ethic—all represent types of consciousness, and they can be altered by choosing a new consciousness."

I believe the message has percolated down through the years. The counter-culture has been successful in bringing about a new consciousness. Later in the book, Chapter 11 says, "Consciousness III will destroy the corporate state 'without violence, without seizure of political power, without overthrow, of any existing group of people.' Counter-culture is sworn to attack minds, not capital gains or depletion allowances. … The hope that counter-culture will transform society into 'something a human being can identify as, home' rests on the fact that it is a middle-class movement. What makes it so important 'is that a radical rejection of science and technological values should appear so close to the center of our society, rather than on the negligible margins. It is the middle-class young who are conducting this politics of consciousness."

For forty plus years, they have continued quietly "conducting this politics of consciousness" with results I suspect the author of the book could have never imagined. As we have seen, they mostly do not hold political office. Without violence, they have attacked minds. They have transformed society—all while we were sleeping (see Section 4, Chapter 12). Toward the end of the chapter, the author, Marvin Harris, again quotes Theodore Roszak, stating that the primary goal of counter-culture is to proclaim "a new heaven and a new earth."

In general, with each of the thirty-plus-year-old quotes, if you replace the words "counter-culture" or "Consciousness III" with "environmentalism," they could have been written today. The environmentalists do definitely dream of a new earth. As you've seen, they have visions of a utopia. They do not have any real plans as to how we will get there. But they are conducting politics of consciousness and, in thirty-plus years, have managed to transform much of society into believing with them.

Patrick Moore is one of the founders of Greenpeace. His experience with the environmental movement of the '60s and '70s supports the thesis of ***Cows, Pigs, Wars and Witches***—though I am not saying he's ever read it. I had the opportunity to hear Patrick Moore when he spoke in New Mexico in 2009, where he said that the environmental movement depends on politics of confrontation—which he defined as telling people what they should not do. His comment was specifically in reference to energy.

Moore explained that the '80s ushered in the age of environmental extremism. The basic issues, for which he and Greenpeace fought, had largely been accomplished, and the general public was in agreement with the primary message. In order for the environmentalists to stay counter-culture, they had to adopt ever more extreme positions. "What happened is environmental extremism," said Moore. "They've abandoned science and logic altogether."

Moore said the extremism movement is dying off because young people are more accepting of new technology. The extremist message today is "anti:" anti-human, anti-science, anti-technology, anti-trade and globalization, anti-business and capitalism, and ultimately, anti-civilization. No wonder it seems no regulation goes far enough to appease them.

Addressing his change in position, Moore says, "I had been against at least three or four things every day of my life for fifteen years; I decided I'd like to be in favor of something for a change. I made the transition from the politics of confrontation to the politics of building consensus." Today, Patrick Moore is known for his work supporting nuclear energy.

To the environmental extremist, as Patrick Moore calls them; the Watermelons, as I call them; or Big Green, as Ron Arnold calls them; America, as we know it, is bad. But if we follow their plan, we will have a new earth—and it will not look much like the country people are risking their lives to get into. It will not be a super-power. Instead it will be a superconciousness, looking more like a hippie commune than the wealthiest nation in the world.

Which of today's "necessities" would you want to live without? As seen in the sidebar, for economist and actor Ben Stein, he would not want to live without the air conditioner. For you, maybe it is not the air conditioner—though probably none of us would really want to live without it. Most of us would have a hard time picking just one modern convenience we would not want to live without. You might think, "I really like hot water when I turn the handle. My cell phone is very convenient. I do like clean clothes. Then there is my car. Oh, my computer!"

Long may it cool
Ben Stein

Let us now praise famous men. Let us praise Willis Haviland Carrier, the main inventor of the greatest gift to mankind since the constitution. Namely, air conditioning. Like most Americans of my generation, in most parts of this nation, as a child I approached summer in my hometown in Maryland with dread. The heat and humidity were unbearable. Sweaty, oppressive days. Torture trying to sleep in painfully drenched-in-perspiration sheets, wet from fever-cruel heat.

Summer car travel was genuine agony. The only place of refuge was a few air-conditioned stores and all movie theaters. I often wonder if movies seemed so much better in those days mostly because of the air conditioning in the theater.

Then, one miraculous day in 1955, my parents bought me a little window Carrier air conditioner. Paradise! I worshipped that machine. Then in '56 my dad got the whole house air-conditioned. The mood in the home was permanently better.

Now, of course, we take air conditioning for granted. Or, maybe not. I thank Mr. Carrier every time I press the buttons, twirl the thermostat, breathe in cool, gorgeous air, take those negative ions into my blood stream, feel protected from the vicious heat of summer.

Now, I know we all have to be careful about energy, but we have to be careful not to die from heat. When I think what air conditioning does, what a miracle it is in terms of human well-being, what a change it has made in human happiness, I am humbled. Take away my Cadillac if you must. Make me wear a sweater and a coat in winter indoors. Make me do anything else, but please leave me with that gift from Mr. Carrier and his colleagues. Long may it cool.

Each year as the Christmas season approaches, inventors come up a new energy-using appliance, mechanical toy, or computer product that we cannot live without. Children must have what their friends have—even if the item is close to lethal or exorbitantly expensive. We all want to be "with it." We plug in one more energy-draining unit, while praising the idea of "going green." It's a patriotic idea, as long as it takes nothing away from our personal pleasures.

The records for 2010 show that Amazon's bestseller was the Kindle—a simple tool that allows us to read books without ever really buying one. No trip to the library or bookstore. We stay warm in our well-heated homes and plug the Kindle into its source of energy.

We are used to available energy. We do not want to be without it, and we don't want to take a big step backward

If you like the freedoms and luxuries that make America great, you, too, can thank Willis Haviland Carrier. But on the bigger scale, you'd better thank energy, as it is what separates us from the undeveloped countries. This is why it is important that we smash the Watermelons—no, not them personally (I am not inciting violence)—but their pervasive hold on America's energy policy. It's energy that makes America great!

Chapter 8

They are Not Who You Think They Are

The environmental movement has been hijacked and the rank-and-file tree hugger is being used, while they cling to the old notion that there are good guys and bad guys—white hats and black hats. Today, it is not so easy to tell who is who.

Yes, as was pointed out by Patrick Moore in the last chapter, there was a time when the dividing line was clear. Big corporations did do nasty things to the environment. Little people threw empty paper cups out the car window. Air and water became polluted and streets were littered. Perhaps some were intentional shortcuts to boost profits. Others were done in ignorance. (When I threw a gum wrapper out the window as a child, I had no malice or willful disregard for the planet.) But, we'd reached a point where the population was such that the earth could no longer hide the abuse. Something had to be done.

The environmental movement was born. Messaging was crafted and people began to see the harm they were doing and changed their ways. America in the 1970s was where China is today. That country is just beginning to work on cleaning up. We've been pretty successful. Ultimately, I trust, China will be, too.

Today, most Americans universally agree on the need for clean water and clean air. Virtually no one wants to strip mine the whole continent or cut down all the trees. Today mining companies receive awards for the masterful, computer-aided, reclamation jobs they do to restore the land when they are finished responsibly extracting the resources. Private forestry is done with planning and care, as new trees are planted to replace those that are harvested. The overall health of the forest is of utmost concern. The same is true for commercial fishing. No one cares more for the resource than those whose livelihood depends on it. Those who used to be thought to wear black hats, now wear white.

Yet, perhaps believing the earth should be back to some prehistoric utopia, environmental groups—Watermelons—attack the very people working to wisely manage the resource.

As we've seen in Section One, giving the Watermelons control is not going to save the planet—and it will ruin America and all of humanity.

We've seen how certain individuals have morphed into the environmentalists of today. But how'd they get so much power? When did they become more than people who want to watch birds, take hikes, and save animals?

The Humane Society of the United States is not what you think

The Washington Examiner, 2 December 2010

Ron Arnold

If you donate to the Humane Society of the United States for supporting the homeless doggies and kitties in your local animal shelter, you've likely been suckered by one of Big Green's most notorious propaganda mills.

HSUS is a radical no-meat, anti-hunting, anti-gun octopus that spends millions swallowing other animal groups whole, but habitually gives less than 1 percent of its annual revenue to a few selected local shelters. It's not that they can't afford it—they raked in $101.6 million last year alone. (2009)

HSUS got so big because of CEO Wayne Pacelle's takeovers of the extremist Fund for Animals and the respectable Doris Day Animal League, in what he calls "corporate combinations."

That, and his ability to create high-profile projects, such as disaster relief crews that swoop stranded cats and dogs from flooded housetops while the TV cameras watch—then quietly dump them on nearby shelters without so much as a dime's worth of support.

Pacelle gets the credit; the grunts get the critters.

In 2008, *The Los Angeles Times* said: Pacelle couldn't get more aggressive without some CGI movie magic.

Most local animal shelters have caught on to him. Many prominently display signs and website disclaimers, reading: "Not affiliated with the Humane Society of the United States."

They're angry that HSUS spent $22.3 million last year siphoning $51.9 million from animal lovers, so they could lobby against "factory farmers," harass hunters and anglers, file lawsuits against circuses and animal acts, and muscle farmers and ranchers into bankruptcy.

Two weeks ago, Brandon Hunnicutt, president of the Nebraska Corn Growers Association, told its annual meeting that HSUS is not out to change agriculture—"they're out to completely destroy it"—the living embodiment of lawyer Christopher Manes' 1992 book, *Green Rage: Radical Environmentalism and the Unmaking of Civilization*.

But HSUS literature and news releases don't give you a clue. Last week, HSUS campaign director Paul Shapiro—who once climbed atop a McDonald's at lunch hour with a "No Meat" banner—told KOLN-TV News in Lincoln, Nebraska, "The HSUS has about 11 million supporters across the U.S. These are supporters that donate to the organization."

That's odd. Supporters who donate get the membership magazine. The HSUS 2009 Form 990 says HSUS sent out its bimonthly membership magazine to 450,000 people—and there are penalties for filing false IRS reports.

But it's a standard Big Green tactic: They lie.

Defenders say HSUS is not a Big Green group, but an animal welfare group far removed from the environmental movement.

Then why is Joshua Reichert, mastermind of Pew Charitable Trusts' Environment Group, on HSUS' board of directors? And Mary Max, trustee of

> Global Green? And Andrew Rowan, president of the Center for Respect for Life & Environment?
>
> And get this: HSUS Treasurer and CFO G. Thomas Waite III is also an officer of ten other related groups, including EarthKind USA, Earth Voice International, and the National Association for Humane and Environmental Education.
>
> Big Green pretenses get completely ridiculous when you see which foundations are giving them how much for what. Like the Marisla Foundation (Getty Oil money) and its $50,000 HSUS grant for "Mexico's regulation of marine noise impacts."
>
> And the Ford Foundation's $185,000 "[t]o optimize the participation of foundation grantees in the World Summit on Sustainable Development."
>
> Then, of course, the Harold Simmons Foundation's $10,000 to HSUS for "Cocktail reception."
>
> This holiday season donate to your local homeless doggy and kitty shelter, not HSUS.

Today, they are not who you think they are. The Birkenstock, love-bead wearers have grown up. They wear suits when needed, and they lobby Washington. Working with big budgets and contributions from wealthy foundations, they have almost unlimited, tax-free dollars to push their agenda. They have cadres of attorneys with the sole goal of stopping things—attacking private property rights and free enterprise—in the name of saving nature. Their efforts are connected—if not coordinated. I am not quite suggesting a conspiracy—though some would go that direction.

Their efforts thwart the American dream as only big companies can afford to comply with the regulatory process and buy off the environmental groups and unions.

Groups such as the Sierra Club, Theodore Roosevelt Conservation Alliance (the word Alliance was changed to Partnership in 2002), Greenpeace and Environmental Defense Fund each started as what we generally think of as an "environmental group." They worked to save the planet by cleaning up pollution and encouraging outdoor activity. They wore the white hats.

Does Big Green really care about wind and solar energy?

The Washington Examiner, 9 December 2010

Ron Arnold

People wondered when Sen. Dianne Feinstein, D-Calif., introduced legislation last year to forbid development in a million-acre swath of the Mohave Desert, curbing thirteen proposed solar plants and wind farms.

But it raises a lot more questions when you see the Center for Biological Diversity team up with California Unions for Reliable Energy to intervene, obstruct, delay, strangle and wreck—no, not another oil rig—a whole string of large solar energy projects.

The Center for Biological Diversity couldn't care less about workers or their unions. They're just glad to have somebody else yelling about endangered species.

But what's in it for job-seeking blue-collar unionists to team up with job-killing greenies?

The 250-megawatt, 1,800-acre Genesis Solar Energy Project got its California Energy Commission license a little over a month ago. CURE fought it alongside the CBD, alarmed about the endangered desert tortoise and American Indian cultural sites.

Who is this CURE? It's a project of the State Building and Construction Trades Council of California, a coalition of fourteen affiliated unions.

CURE has a long track record of playing environmentalist—the energy commission told me CURE has intervened to stop eight of the nine solar energy developments on this year's docket, and old newspaper clips show CURE doing the same for many years.

Why? To choke developers with impossible demands until they accept a project labor agreement, or PLA, a controversial package deal that gives a ruling union council great power—to allow only its selected unions to work on projects, to make nonunion workers join a local union, and charge 25 cents per hour for each employee for its war chest, in exchange for a guarantee that workers won't strike.

Just sign on the dotted line, and suddenly your foe becomes your friend. All those endangered tortoises and disturbed burial sites are forgotten.

But wait, isn't that blackmail? Developers have called it "greenmail" and "environmental extortion."

Not every union approves of CURE's tactics. When Genesis Energy went through its decisive hearing in August, Danny Curtin, director of the California Conference of Carpenters, listened to hours of CURE's complaints, then got up and said to the commissioners:

"Let the project stand or fall on what it needs to deliver. Move off of all of the things you've heard from the CURE group about this environmental issue or that environmental issue. The motivation behind that is really inappropriate."

> Just to make sure the commission got the message, Douglas J. McCarron, general president of the United Brotherhood of Carpenters and Joiners of America, sent the commissioners a letter:
>
> "I urge you to be especially cautious of any labor organizations that express opposition to these urgently needed renewable-energy projects on environmental grounds. While it is their right, their motives may be to pressure developers and their chosen contractors to sign outdated labor agreements that do not fit the needs of this newly emerging and evolving industry. It will not only add extra, unnecessary, uncompetitive costs, it will hamper and obstruct the process of innovation essential to this vital industry. The clean energy revolution is too important to allow these tactics especially under the guise of environmental concern."
>
> The California Energy Commission took McCarron's advice and ignored CURE and CBD this time, ending the process in time to make Genesis eligible for federal stimulus cash grants.
>
> Genesis got its license and negotiated a satisfactory union-backed labor agreement. It doesn't always end this way, but Commissioner Robert B. Weisenmiller said the state's need for renewable energy and the 650 jobs Genesis would create outweigh the downside.

Somewhere along the way, they got, as Patrick Moore calls it, "hijacked." The dates are different for the diverse groups. For example, the Environmental Defense Fund was started in 1967 by a group of lawyers and scientists. They were a single-purpose group whose goal was to ban the insecticide DDT. With lawyers at the core, they set out to "defend the environment in the nation's courts." They failed in their attempt to ban DDT in their home county of Suffolk County in New York but did not give up their passion. They'd discovered the power of the courts, which gave them greater clout toward saving the environment than the meek "conservationists" were exerting.

Long Island Attorney Victor Yannacone was one of the original founders. He left the EDF in 1969 because they had become a perversion of their original purpose. Additionally in 2008, Yannacone issued a controversial argument against the

EDF's stand on man-made global warming. His "open letter to honorable Journalists" concludes with this:

> "The real climate change 'deniers' are those who refuse to acknowledge the correlation of earth temperature with solar activity. This correlation can be back modeled and is a useful forecasting tool. There is no correlation between carbon dioxide and earth temperature. Current computer models cannot back-model temperature history and thus are useless for forecasting. Yet the mass media and many concerned citizens who should know better continue to prattle on about human industrial activity causing global climate change?
>
> "In the coming colder earth, we are going to need all the energy we can muster to keep poor nations and poor people from starving and freezing. The only solution to global warming, according to those seeking quick profits from mass hysteria, is to reduce the supply of energy in the developed world. If, as the same pundits predict, we will have to accommodate a nine-billion-person world in less than fifty years, how will we feed all those people?"

Likewise, as we've already addressed, Patrick Moore, one of the founders of Greenpeace back in 1971, left the organization in 1986 because they had become too extreme.

The Theodore Roosevelt Conservation Partnership (originally Alliance) was started in 2000 as a part of Trout Unlimited (TU). Their stated purpose is to represent "a coalition of the nation's leading hunting, fishing, and conservation groups working together to guarantee all Americans quality places to hunt and fish." In 2002, their status with TU changed, and they became an independent organization. The name changed, new leadership was hired, and new donors came on board. In 2005, the Theodore Roosevelt Conservation Partnership (TRCP) received a $600,000 grant from the Gordon and Betty Moore Foundation with the specific mandate "to change the current course of energy development on public lands." TRCP's work began to change with the initiative. They launched a barrage of attacks on the oil and gas industry—despite the fact that the two can peacefully co-exist. Well, they can co-exist.

Then we come to the Sierra Club. Probably the oldest of the environmental groups, founded in 1892—long before the term "environmentalist" was in the dictionary. Because of their long history, they are a good example of the environmental movement's evolution from true naturalists to today's environmental extremists.

Founded by a number of California notables, including naturalist John Muir and University of California paleontologist Joseph LeConte, The Sierra Club remained a local hiking group of the San Francisco elite for many years. Muir had advocated wilderness protection for the Sierra Nevada through government ownership since the 1870s and was influential in establishing Yosemite National Park in 1890. After the Yosemite success, Muir's supporters in the East immediately began thinking about an organization to protect the remainder of the Sierras. Simultaneously, students and faculty at the University of California in Berkeley were planning an alpine club. The two factions got together and, on June 4, 1892, twenty-seven men signed the articles of incorporation in San Francisco. Muir served as its first president.

Never let the truth stand in the way
(24 April 2009)

When we were children, my mother gave the family humorous retellings of the day's events each night. In response, we had a saying, "Never let the truth stand in the way of a good story." Her creativeness was entertaining. It wasn't mean and didn't hurt anybody.

I was recently reminded of the epitaph during the State Game Commission hearing in Farmington, NM. Overall the hearing was positive. The Commission did a great job seeking input from the stakeholders and fishing groups.

The general tone of the meeting made the comments of two known dissenters stand out. In contrast, their vitriol flowed freely.

Tweeti Blancett said that her ranch is a "Federal Lands Ranch" and that "every one of our wells is contaminated." She claims that the very same water was going into the San Juan and Animas Rivers.

Later, former commissioner Oscar Simpson spoke. He repeated his usual rant about how oil and gas runoff is a long-term problem that is exacerbating deterioration and that fishing and hunting are declining due to lack of regulation. This is the same Oscar Simpson, who was responsible for the $10,000 radio ad that aired in the summer of 2008 in Albuquerque, implying the oil and gas industry ruined fishing on the San Juan River due to oil-soaked debris.

Both blamed the industry for damage to the water quality of the San Juan.

During the meeting Steve Henke, then District Manager of the Farmington BLM office, spoke. With regard to sedimentation of the quality waters, he referenced a study by the United States Geological Survey of the Largo Canyon Watershed. He indicated USGS found that the naturally occurring erosion was the dominant contributor, and that any additions from oil and gas development were "insignificant." He stated the sediment entering the San Juan at Simon Canyon was also predominately associated with the natural-occurring events, similar to the results of the Largo Canyon study. Mr. Henke added that the New Mexico Environmental Department Surface Water Control Bureau has designated both the Animas River, and the quality water section of the San Juan River as attaining water quality standards for designated uses, meaning they are not impaired.

Mike Sloan, Chief of Fisheries, New Mexico Department of Game and Fish, reported that if hydrocarbons were present in the river, they'd be so toxic to the fish that they'd see mortality. But the fish are healthy and thriving.

Rudy Rios, President of New Mexico Trout, reported on a survey of his members. They were asked about water and fish quality, and sediment. The response to each was overwhelmingly positive.

The others disputed Blancett and Simpson's claims of "damage."

I suggested the comments of Tweeti and Oscar had to be measured by the well-known fact that they have made it their life's work to destroy the oil and gas industry—regardless of the truth. I've followed Oscar's efforts since I'd been alerted to them by the 2008 radio ad, and I'd been introduced to Tweeti through her efforts to block oil and gas production in Santa Fe County. I told the Commission that this was not really about the river. The San Juan was just the vehicle for their vendetta.

When there was a break in the meeting, people I'd never met congratulated me. Apparently I'd said what they all knew to be true, but had been afraid to say.

I was told that when Tweeti first started making the accusations about water contamination, she was asked to prove her point and provide documentation, which has not been produced. Yet she continues to make such claims. If this is incorrect, I invite her to make the data public. She never has. The typical modus operandi.

The San Juan River Guide Association is outraged at Oscar's false claims about the river. Larry Johnson of Soaring Eagle Lodge handed me a response they'd written about the disparaging remarks Oscar made about the river's condition in an Associated Press article published nationwide. "We do not support or agree with..., we see no facts or data to support many of the conclusions..." Larry specifically said, "We earn our living off the river, if there was any damage, we'd be the ones making noise!" Various members reported hundreds of thousands of dollars of lost business due directly to Oscar's efforts. The San Juan Foundation is working to rebuild the river's reputation.

Though these oil-and-gas opponents attack their villains with vigor, those listening to their vile diatribes need to know that they do not let the truth stand in the way of their agenda.

In 1908, the legendary battle that became known as Hetch Hetchy was waged by the Sierra Club. The goal was to prevent the construction of a water-supply dam in Hetch Hetchy Valley near Yosemite and was not typical of the early Club. The Hetch Hetchy Reservoir was filled. It still supplies domestic and irrigation water—although environmentalists, seeking Muir's revenge, would still like to have it torn down.

The Hetch Hetchy battle is representative of the essence of the preservationist philosophy of John Muir, as opposed to the conservation philosophy of Gifford Pinchot, humanitarian, utilitarian, and first Chief of the U.S. Forest Service. Pinchot coined the phrase, "Conservation is the wise use of resources." Muir called for an end to commodity use of natural resources, a ban to be enforced through government ownership and regulation. Defeat of the preservationist philosophy left the Club a political moderate for many years.

In 1946, the Club had 3,500 members and was restricted to California. However, in 1947, it began to stir politically again, and reversed its earlier policy favoring roads into the Sierras, removing from its bylaws the clause "to render accessible the mountain regions"—a phrase included in the bylaws specifically by John Muir because access was important to him. The Sierra Club had been drifting leftward, but as late as 1949, Club president Bestor Robinson could be heard saying, "Never let your love of nature overshadow your concern for human needs. I want wilderness to contribute to the American way of life."

New voices in the 1950s pushed the Sierra Club closer to its present position. Leaders criticized recreational developments in the national parks and began to sound the theme of exclusion. They clashed with the Forest Service over logging in the national forests and over wilderness protection in the North Cascades. In 1956, the Club joined with the Wilderness Society to seek wilderness legislation rather than administrative protections, resulting in the Wilderness Act of 1964.

David Ross Brower, mountain climber in his youth, activist in his maturity, became the Sierra Club's first executive director in 1952. Brower was a dynamic leader and the epitome of the amateur naturalist as head of a conservation group. It was under his direction that the Club became a militant national organization. Brower was the first, but not the last, environmental leader to relish being called an "extremist." Extremism to Brower meant distrust of authority, public attacks on misguided expertise, and avoidance of compromise.

Michael McCloskey, a lawyer, took Brower's place, ending the era of Club leadership by amateur naturalists and beginning the professionalization of the environmental movement. He revived the bitter images of John Muir's defeat at the hands of Gifford Pinchot in the Hetch Hetchy dispute. It became the Club's "Remember the Alamo!" battle cry. McCloskey's leadership achievements include extensive victories in pollution control, wildlife, and energy. Mike McCloskey was arguably the best star pilot in the galaxy of environmental leaders and one of the most cunning warriors the movement has ever had. Perhaps the most important lesson we can learn from McCloskey's list of accomplishments is that the growth of environmental legislation has systematically prevented new economic development and burdened existing economic enterprises for more than two decades and that no one has calculated—or even thought about—the cumulative economic impact.

While this history of the Sierra Club (adapted by permission from ***Trashing the Economy*** by Alan Gotlieb and Ron Arnold) is extremely abbreviated and, for the sake of brevity, omits important developments such as the Sierra Club Foundation and the Sierra Club Legal Defense Fund (remember the earlier comment about cadres of attorneys), its inclusion here is intended to show how a well-intentioned hiking group can evolve into a group of radical extremists—without the rank-and-file members ever knowing the change took place. Their story is representative of many of the Watermelons using unlimited pots of money and lawyers with the goal of locking up more and more land—meaning fewer and fewer can benefit from it.

With many Sierra Club leaders holding urban planning degrees, it's not hard to see why the environmental movement is so in love with centralized planning as a model for reshaping society. Even with the Soviet Union's collapse, they are not about to give up the dream, which led the late newspaper columnist Warren T. Brookes to call members of the Green Movement "Watermelons:" Green on the outside, Red on the inside. There's only one difference: Even old line Communists think the enviros are crazy. Lev Timofeyev, author of ***Russia's Secret Rulers***, once said that environmentalists startle even hard-line right-wing Communist Party stalwarts with their blind faith in planning and government control.

The modern era of the Sierra Club started with the current Executive Chairman Carl Pope in 1992, who had begun shifting the Sierra Club (Big Green) into a formal alliance with big labor just before the Kyoto climate summit in late 1997, when stark calculations predicted massive job losses necessary to cut CO_2 emissions. (Remember, the Sierra Club's roots are that of a wilderness preservation group. "Big labor" doesn't work in wilderness areas—nor does anyone else. Wilderness, by law, is an area where "man is a visitor who does not remain." Development of any kind is strictly forbidden. No homes or stores. No structures of any kind. Roads are not permitted. Mechanized equipment may not enter.) Along with elements from both the blue-collar world and the green-dollar community, The Blue Green working group was created in 1960—initiated by SEIU's Andy Stern. Pope and other enviro leaders kept up a dialogue with key unions. For years the Blues and Greens tried to converge. While the plans finally failed when the September 11 attacks disrupted the meeting of Carl Pope and Dan Becker (the Sierra Club's director of global warming and energy programs), John Adams and Robert Kennedy, Jr., of the Natural Resources Defense Council, John Podesta of Center for American Progress, and John Sweeney, president of AFL-CIO, a link had been formed between two of the progressive movement's most important constituencies.

Throughout this book, you've seen references to the big budgets with which the Big Green groups have to accomplish their missions and fund the legal campaigns. None of this comes cheap. This is another link that connects many of them, one to another: wealthy foundations.

In the brief overview history presented earlier in this chapter, I mentioned a grant to TRCP from the Gordon E. and Betty I. Moore Foundation. At the time that grant was received, John M. Seidl was a "Director" at TCRP. Worthy of note is the fact that he had served as Chief Environment Program Officer for the foundation. They represent just one small part of wealthy donors who stay in the cities and fund these activities.

The listing of grants given from the Moore Foundation in 2007-2008 is twenty-eight pages long. Just to give you a sense of the interconnectedness, here is a small sample of their support in 2008 to groups I've mentioned.

- Trout Unlimited (TU), Arlington, VA, received $76,649 in 2008 for hatchery reform in the Columbia River basin, $653,300 for an Alaska salmon program, and $8,000 for an annual gala.
- Environmental Defense Fund, New York, NY, received $1,362,586 in 2008 for national policy framework for advancing dedicated access privilege programs, $731,331 for New England catch share management programs, and $50,000 for a workshop on climate change in the Amazon River Basin.

These figures do not include millions of dollars given to other groups such as the World Wildlife Fund, the Nature Conservancy, and the Conservation International Foundation.

The Gordon E. and Betty I. Moore Foundation is a lesser-known player compared to the work of George Soros.

Glenn Beck has dubbed George Soros "Spooky Dude." While some scoff at Beck's entertaining presentation of the facts, George Soros' impact on the activities of Big Green is truly "spooky." Soros' Open Society Institute (OSI) gave millions in 2008 (according to their 990 tax forms) to environmental groups—including the World Wildlife Fund, which also received funds from the Moore Foundation. While many contributions are listed as for "general support," the largest single donation to an environmental group is for $5,000,000 for the Alliance for Climate Protection—the group that paid for the television commercial featuring Nancy Pelosi and Newt Gingrich together supporting fast action on climate change. (Interestingly, the Alliance for Climate Protection does not list OSI as one of their "Partners" but they do list another OSI grant recipient, Green for All, and the Blue Green Alliance mentioned earlier.)

There are also specific gifts from OSI to the Natural Resources Defense Council "to train and mentor lawyers interested in pursuing environmental litigation cases through a two-year fellowship program." In previous years, OSI has provided grants to many other groups, including the Environmental Law Institute "to provide support for the Endangered Environmental Laws Project." (Remember the note about funding legal campaigns?)

When you "follow the money," you find that several other foundations appear repeatedly, just in the groups I've mentioned here. These foundations include the Joyce Foundation, Rockefeller Brothers Fund, Tides Foundation, and the Turner Foundation.

The data presented here is just a snapshot intended to give you a glimpse into the work of the Watermelons—or Big Green. Whether you call the environmental groups by one moniker or another is not the point. These groups have given themselves names that mask much of their true intent: red on the inside.

Throughout history there have always been those who seek ultimate power. Germany's Hitler and Japan's Tojo are some of the most notable. Most of us are good people who are naïve when it comes to the psychology of the control freak. We cannot even imagine that there are really people who set out to destroy and control. These control freaks are using the well-intentioned environmentalists to gain power and diminish American sovereignty.

The environmentalists are being used as a tool to help collapse America. If they succeed, what would it look like?

As we've done in Section One, if we draw the Watermelons' actions out to their logical end, would America be a better place; would the world? No.

Who stopped the control freaks of the past? America. How did these past regimes deal with national treasures? Hitler tore them down en route to accessing the natural resources—oil and gas—in Romania and Russia.

The next generation of global leaders will make Hitler's efforts look like child's play. They will plunder the resources on their way to totalitarian control. Which countries that are under a dictatorship or communist control protect the environment? These groups do not care any more about the environment than the Chinese do.

Most of us, me included, believe we are "environmentalists." We agree on a lot. We all want to protect the special places—the rocks and ruins. We want clean air and a safe food supply.

We also love America and understand that America needs to be strong to protect the less fortunate—including the environment.

Environmentalists—not Watermelons or Big Green—please consider what would happen to the natural resources if the progressives win. They are not for "green." They are for bringing down America. We can't go back. We need energy and we need it responsibly. After all, energy makes America Great!

Chapter 9

Rhetoric: The Power of Language to Manipulate Opinions

By Lawrence J. Clark

In the heated political environment, the dreaded words "Hitler" and "Nazi" are frequently tossed about. More than the accusation some see on the surface, their invocation is more about a comparison of tactic—specifically the idea that if one tells a lie often enough, people will believe it. This tactic is known as "rhetoric"—or using language to persuade.

Rhetoric was a favorite tool of Hitler and the Nazi movement. The same techniques that were successful then, were used before, and are still in use today. The Greek philosopher Aristotle defined rhetoric as "the ability, in each particular case, to see the available means of persuasion." He narrowed the major "available means," or methods, of persuasion down to three parts: *ethos* (the character or perceived authority of the speaker/writer), *pathos* (manipulating the emotions of the listener/reader) and *logos* (using facts, statistics, logic, etc. to support the main argument).

History tells us that the Germans fell prey to Hitler's persuasive powers that ultimately led to their downfall. Through this chapter, you'll see how rhetoric is still being used and how, if we are not careful, we, too, will fall prey to "nice" sounding ideas that can lead to our peril.

Before we can solve the problem, we have to be aware that there is one.

Rhetoric is widely used in political campaigns and in advertising. When we hear a politician tell us he or she will fix all the current ailments, most of us know to "take it with a grain of salt." We know to check their voting record and personal history. We accept that they are full of bluster—after all, he or she is a politician. When we see or hear an advertisement for a product that will "slice, dice, and chop," we generally know that it is "too good to be true." We recognize the work of Madison Avenue and know the overstatements are part of the game.

> If you tell a lie big enough and keep repeating it, people will eventually come to believe it. The lie can be maintained only for such time as the State can shield the people from the political, economic and/or military consequences of the lie. It thus becomes vitally important for the State to use all of its powers to repress dissent, for the truth is the mortal enemy of the lie, and thus by extension, the truth is the greatest enemy of the State.[83]
>
> Joseph Goebbels, Hitler's Minister of Propaganda

Where most people are more gullible (especially those who've not read this book), is regarding the "save the earth" messages of the "well-meaning" environmentalists. Somehow we believe that they are sacrificing their potential financial success for a noble cause. Their messages about an endangered species such as the spotted owl, abused puppy, or a threatened landscape generally escape the typical public skepticism. Yet, like those throughout history and the modern day politician or Madison Avenue advertising copywriter, the environmental groups are aware of the power of propaganda.

The use of propaganda is not limited to the use of manipulating Americans' needs and desires to persuade us to buy products; it is carried on through fundraising efforts and environmental groups. "What you get in your mailbox," admitted National Audubon Society Chief Operating Officer Dan Beard, "is a never-ending stream of crisis-related, shrill material designed to evoke emotions, so that you will sit down and write a check."[85]

> The conscious and intelligent manipulation of the organized habits and opinions of the masses is an important element in democratic society. Those who manipulate this unseen mechanism of society constitute an invisible government, which is the true ruling power of our country.
>
> We are governed, our minds are molded, our tastes formed, our ideas suggested, largely by men we have never heard of. ...it remains a fact that in almost every act of our daily lives, whether in the sphere of politics or business, in our social conduct or our ethical thinking, we are dominated by the relatively small number of persons... who understand the mental processes and social patterns of the masses. It is they who pull the wires, which control the public mind, who harness old social forces and contrive new ways to bind and guide the world.[84]
>
> Edward Bernays, ***Propaganda***, 1928

Despite the staggering cost of producing hundreds of thousands of slick, four-color inserts, the trees that were killed to produce the paper (or the birds and furry little squirrels who would have lived in those trees), the pollution generated by manufacturing the ink or the little plastic see-through windows, or the CO_2 emitted into the atmosphere by all of the U.S. Postal Service vehicles roaming the streets to deliver the expensively produced solicitation packets, the rhetoric works. Sierra Club conservation director Bruce Hamilton understands the implications, but has a "well, the show must go on" attitude. "I'm somewhat offended by it myself," he admits, "both intellectually and from an environmental standpoint. And yet, it is what works. It is what builds the Sierra Club."[86]

And build it they must, as the organization loses about 25 percent of its membership each year through non-renewals, and only about one of every hundred solicitations results in a donation. This is why most fundraising letters from these groups begin with the prediction of some catastrophic result unless "X" amount of dollars is raised within the next "Y" period of time. Without such dramatic (albeit alarmist and often misleading) proclamations, the return on the advertising/fundraising expenditures would be even lower.

Next time you get such propaganda in the mail or see an advertisement on television, watch what they are doing and pay attention to how they are doing it.

Ethos: The Credible and the Incredible

One way *Ethos* is used is to persuade the public is by lining up endless celebrity endorsements. We've all seen them; the latest movie starlet or pop singer du jour devotes part of her Golden Globe or MTV video award acceptance speech to publicize her pet cause. Pop star Jason Mraz showed up at the Grammy Awards Show[87] wearing an "orglamic" eco tux made of 100 percent recyclable fabric created from recycled PET (Polyethylene terephthalate or plastic) bottles.[88]

Chris Carter was the co-writer, co-producer, and director of the popular television series, *The X-Files.* In 2008, Carter directed a second feature film based on the series, *The X-Files: I Want to Believe.* The DVD's bonus section contains a six-minute feature in which Carter proudly explains how the film was produced using "green production" techniques: all vehicles were Ford hybrids, the catering companies used locally produced food and purchased only from businesses with good sustainability practices, and all the sets were recyclable. Joshua Zyber, a reviewer at High-Def Digest called the effort "sanctimonious." "Some sanctimonious preaching from the director about how he is single-handedly saving the planet by recycling water bottles on set and using hybrid cars to drive the cast and crew around."[89]

Another way the environmental groups use and abuse ethos is by quoting or using scientists, doctors, and other authoritative/credible figures to promote their causes.

The National Geographic Society had a public service announcement on CNN. The spot features a female "scientist" with a Ph.D. who claims that the earth is running out of water (not fresh water, but ALL water). Since 70 percent of the earth is covered with water, and the average depth of the ocean is about 3,000 feet, even a lowly English major like me finds this claim hard to believe. Unfortunately, since the scientist seems so sincere and authoritative with her credentials flashed on the screen as she speaks, and is sponsored by a well-known and "trusted" organization such as National Geographic, many people probably take her word for it.

Another advertisement I saw was cleverly disguised as a news story, "Global Warming: Dire Prediction for the Year 3000," which appeared on livescience.com and was widely distributed by Yahoo News in early 2011.[90] Although we will discuss the article more fully in our section on the use of pathos, or playing on people's emotions, it is also a good example of floating an idea in the public via an authoritative person (associate professor and his team of researchers at a university, who is using a credible method of inquiry [computer modeling]). Their main hypothesis is that even if humans stop creating excess carbon dioxide in the next hundred years, the "lingering effects" of global warming will cause global sea levels to rise thirteen feet and the West Antarctic ice sheet to collapse.

Pathos: Playing with Emotions

Pathos is the rhetorical strategy of playing on people's emotions in an attempt to persuade. Whether you are familiar with the term or not, you have most definitely seen this strategy in action. Buy this beer, and you'll be surrounded by beautiful women (and therefore happy). Sign up for our dating service and meet the man of your dreams (and be happy AND secure AND personally fulfilled). Buy this suit and impress your boss and coworkers (and be happy AND secure AND personally fulfilled AND confident AND rich enough to afford the beer in the first commercial and attract the woman in the second).

As much as happiness sells, *FEAR* is the emotion most played upon by environmental activist fundraisers and propagandists. "There will be no polar ice by 2060," says Larry Schweiger, president of the National Wildlife Federation. "Somewhere along that path, the polar bear drops out."[91]

This statistic was used in a YouTube video in which faces of high school students fade in and out, as in the background we hear the ticking of a clock that speeds up gradually as the video progresses. The video leads viewers to a website called fightglobalwarming.com, which is sponsored by the Environmental Defense Fund and the Robertson Foundation.[92] On the "About Us" section of the website, the Environmental Defense Fund uses pathos to create a sense of urgency and stir up fear in the reader (italics added):

> We've launched this campaign as a *wake-up call.* Global warming is the *most serious environmental challenge* of our time. It is *more urgent* and its *dangers* are more fundamental than most Americans realize. This campaign seeks to educate Americans about *how quickly we must act* . . .

The Robertson Foundation takes the emotional language even a step further:

> The Foundation believes that we have an obligation to be responsible stewards for future generations. In that regard, we believe that *failure* to address global warming *imperils* future generations by *saddling them* with lasting and potentially *devastating consequences.*

Is there no stopping the seemingly never-ending flow of information from the global warming/global climate change propagandists? Following is an excerpt from the livescience.com article mentioned in the *ethos* section of this chapter:

> Even if humans stop producing excess carbon dioxide in 2100, the lingering effects of global warming could span the next millennia. The results? By the year 3000, global warming would be more than a hot topic—the West Antarctic ice sheet could collapse, and global sea levels

> would rise by about thirteen feet (four meters), according to a new study.
>
> Using a computer model, researchers looked at two scenarios—an end to humans' industrial carbon dioxide emissions by 2010 and by 2100—stretched out to the year 3000.

Even if humans were to stop emitting excess carbon dioxide—or if they figured out a way to completely capture it—the effects of global warming would continue to accumulate. That's because previously emitted carbon dioxide lingers in the atmosphere and the oceans, unlike land, warm only gradually, according to one of the study researchers, Shawn Marshall, an associate professor of geography at the University of Calgary.

Talk about invoking mass hysteria! My dad, who spent his career as a meteorologist with the U.S. Air Force, constantly reminded us kids (and the TV weatherman) that forecasting the weather is mostly guesswork—educated guesswork, but guesswork nonetheless. This kind of story reminds me of hearing politicians telling us that a certain program or tax break is going to cost X billion dollars (or only X billion dollars), only to find out that the number is spread out over ten or twenty or even thirty years, and/or is based on estimates of the inflation rate or interest rates or some other highly unpredictable statistic. And of course there is no mention of the fact that without atmospheric CO_2, Mother Earth would be unable to produce such necessary items as grass, trees, vegetables, etc. and that CO_2 was present in the earth's atmosphere long before the invention of the internal combustion engine.

Yet another example of the rhetorical strategy of playing on people's emotions—in this case FEAR—the foolproof and effective fundraising tool. Look for quotations from the above article to appear in a fundraising letter in your mailbox soon!

Logos: Just the Facts

Logos is the rhetorical strategy of using logic, facts, statistics, etc. to persuade the audience. In the case of the messages sent out from big green, the "facts" can be made up or quoted as settled science, when it is far from it. Nonetheless, the information is presented as if it is established fact and stated, as previously discussed, with authority.

A television spot from Platform Breathe displays this. While a cute little girl draws a picture in the sand, we hear a mellow bluegrass tune in the background underneath the following voice over:

> This is Joe. He really likes the sun. Joe says the sun can give us all the energy we need without hurting the planet. He says there are people and companies who already know how to use the sun. We just have to get behind them. Can one solution to energy woes be found up in the sky? The possibilities are endless.[93]

Joe is a face drawn in the sand that will wash away with the next tide (powered by the moon's gravitational pull on the earth, not by fomenting seas or tidal waves created by a mini-van's exhaust pipes). Joe delivers no hard statistics but the script is delivered as if it were fact—and the statements are made by a cute little girl who plays on our emotions. It has enough of what we generally believe to be fact to accept it all hook, line, and sinker.

But think about it. Joe "says there are people and companies who already know how to use the sun." We know this to be fact. Most of us have some solar-powered items in our life—maybe a sidewalk light or a water heater. Solar power is used in the oil fields. We all know of cases where commercial rooftops are lined with solar panels. Yes, there are "people and companies who already know how to use the sun."

> The romance of renewable energy is such that we are ignoring logic and common sense, as well as hard facts and figures.
>
> Robert Bryce, Author, ***Power Hungry***

"Joe really likes the sun." This is not disputable one way or the other. Joe is a smiley face drawn in the sand—in the sand on the beach. We can accept this without much question.

"Joe says the sun can give us all the energy we need without hurting the planet." In this statement, there are several questionable elements and some flagrant inaccuracies. First let's look at "hurting the planet." This implies that all other energy sources do hurt the planet—which is not fact, even if you believe in manmade global warming. For example, hydropower emits no emissions. If you want to get very technical, a mill can be powered by the natural flow of a river. The waterwheel can be made of wood and crafted without power tools. If it is emissions you feel hurt the planet, nuclear is good. Perhaps, to you, "hurting the planet" means scarring the earth—such as mining for coal. What about the steel, copper, quartzite gravel, silicon, titanium, phosphorus, lead, aluminum, and rare earths used to make the photovoltaic panels?[94] All of these ingredients are mined and they require energy to turn them into something that can convert the sun's energy into electricity.

Additionally, solar power requires massive land use and offers comparatively little payback in the form of energy. Large-scale concentrating solar and solar photovoltaic power plants require approximately 6 acres of land per megawatt of electricity output, while nuclear, natural gas, and coal power plants all require about one-half to one acre of land per megawatt of electricity output. Typical commercial scale power plants are 100 to 500 megawatts in size, with often one to four plants at a given site. In Albuquerque, the local utility company, PNM, reports that their second-largest electricity user is a manufacturer of solar panels. Wouldn't you think that someone who makes solar panels could just cover the factory's roof and be self-sustaining?

Plant Type	Plant Size (MWh)	Land Area (acres)	Land Size	Unuseable Land Size	Capacity Factor
Coal/ biomass or gasification w/ steam turbine	500 - 1000 MW	640	1 mi x 1mi	All	.95
Nuclear Steam	500 - 1000 MW	640	1 mi x 1 mi	All	.95
Natural Gas Combined-Cycle	200 -500 MW	320	0.7 mi x 0.7 mi	All	.95
Geothermal Steam	200-500 MW	320	0.7 mi x 0.7 mi	All	.95
Concentrating Solar	500 MW	3000	2.2 mi x 2.2 mi	All	0.3 - 0.4
	1000 MW	6000	2.8 mi x 2.8 mi	All	
Wind	500 MW	23000	6 mi x 6 mi	640 acres	0.3 - 0.4
	1000 MW	46000	8.5 mi x 8.5 mi	1280 acres	

95

Next, look at "all the energy we need." We need a lot of energy. For example, the Energy Information Administration, Annual Energy Outlook 2011, forecasts a 25% increase in US electricity generation growth between 2010 and 2035—1% growth per year. In 2010, in the United States, we used 4,120 billion kwh, in 2035 we'll need 5,167 billion kwh. Between 2010 and 2035, the transportation sector grows by 16.5% from 27.42 quads in 2010 to 31.95 quads in 2035, an annual growth rate of 0.6% a year. "Energy" encompasses both electricity and liquid fuels for transportation. Regardless of the number of gallons of gas or kilowatts we use, the sun only shines in the daytime. To use the sun's "energy" at night, we need some way to store the energy, like batteries, and the batteries are made up of all kinds of elements that are dug up from the earth, such as lead, sulfuric acid, and nickel cadmium.

Saying that the sun can provide "all the energy we need" is, currently, a lie. But wrapped in the pathos of the little girl, said with authority (ethos), and (logos) sprinkled with enough fact to be misleading, the average viewer would believe that all it will take is to "get behind" it.

"If you tell a lie big enough and keep repeating it, people will eventually come to believe it."

Abraham Lincoln said, "You can fool some of the people all of the time, and all of the people some of the time, but you cannot fool all of the people all of the time." Some people may be fooled some of the time by the use of rhetoric in the save the earth campaigns, but not you! You now know how to read, or listen, beyond the obvious. You won't be suckered into writing a check to save the polar bears—or whatever the current scare might be.

You now understand how to spot the rhetorical strategies (ethos, pathos, and logos) that are being used to persuade the general public.

Section Three

Green Schemes

Chapter 10

The Spotted Owl

It is hard to imagine a bigger failure—or a greater success—depending upon which side of the issue you stand. If you strive for open and honest government policy that is straightforward about its goals, this twenty-year experiment has failed. If you believe the end justifies the means, regardless of the cost in life or livelihood, then the spotted owl represents a great success.

Twenty-plus years ago, on June 26, 1990, the spotted owl was listed as "endangered"—an effort that began back in 1968.

The Defenders of Wildlife website states that, "Like all spotted owls, the northern spotted owl lives in old-growth forests." Addressing their behavior, "Northern spotted owls are very territorial and intolerant of habitat disturbance. They prefer old-growth forests with tree canopies that are high and open enough for the owls to fly between and underneath the trees. Preferred areas have large trees with broken tops, deformed limbs or large holes used as nesting sites. Each pair needs a large amount of land for hunting and nesting, and although they do not migrate, spotted owls may shift their ranges in response to seasonal changes, such as heavy snows, that make hunting difficult." The site lists the "loss of old-growth forest habitat as a result of logging and forest fragmentation" as "the biggest threat."

As a result of this belief, the spotted owl was listed as "threatened" and then in 1992: "endangered."

Environmental groups such as the Center for Biological Diversity and the Western Environmental Law Center have used the designation to block logging, cattle grazing, urban sprawl, and power lines—any activities that could "remove forest cover." Their efforts have been successful.

Activities, central to American economic might, have been blocked.

For example, in Oregon, in 1989, logging on federal lands accounted for more than half of Oregon's harvest. As of 2008, it had fallen to less than 10 percent. While some logging exists on private and tribal lands, that, too, is threatened—even endangered. Under the Endangered Species Act, the habitat of the spotted owl is so regulated that private landowners worry about making their timber "too inviting"—as the spotted owl doesn't understand the boundaries of the protected land versus the private. If a spotted owl should be spotted on private land, the owner will be inviting regulation and scrutiny. Private landowners' property rights can be locked down to the point that they are unable to commercially thin their own woods.

The spotted owl dispute has shut down a substantial part of federal timber harvest and threatens logging on private lands. Sawmills have closed—as many as 187 in Oregon, Washington and California, and nearly 100,000 jobs have been lost. Even Eric Forsman, who started the spotted owl saga back in 1968, admits he hates to see the lost jobs, but states: "I feel like it is my job to be a spokesperson for wildlife."

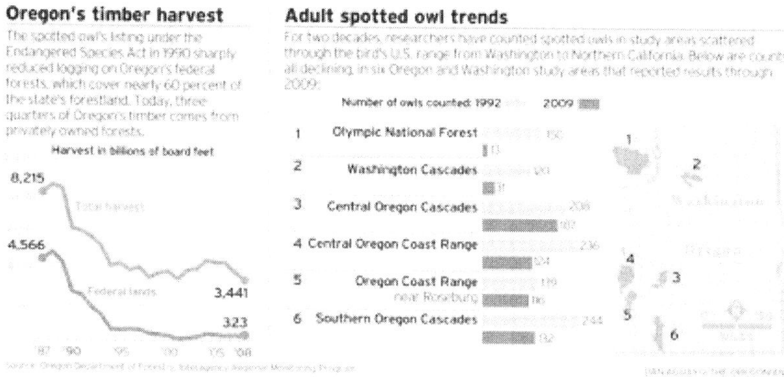

Back in 1994, shortly after the spotted owl was listed as endangered and the habitat protected, Washington Senator Slade Groton said, "The U.S. government, driven by sophisticated, well-financed national environmental organizations and supported by the media and urban opinion leaders, has betrayed rural communities and destroyed—yes, destroyed—the lives and careers of tens of thousands of honest working families in the Pacific Northwest." His comment could have easily been made today. And, we wonder what happened to jobs in America.

Small operators cannot stay in business, as environmental restrictions and legal battles have made logging too expensive for independents. The timber industry is now primarily limited to big companies, which own their own timberlands, and to foreign companies logging in countries without restrictions. (Who do you think logs more responsibly?)

A contact in a small, independent forest-products company in New Mexico told me that when the housing industry comes back, people will be shocked as the wood needed will be much more expensive. Stockpiles have been used up, and there is little new wood available in the U.S. It will all come from other countries.

In 1994, the Clinton Administration introduced the Northwest Forest Plan that was supposed to guarantee specific amounts of logging, but according to Jim Geisinger, executive vice president of the Associated Oregon Loggers, those numbers never materialized. The federal forests were left more vulnerable to catastrophic fires—which hurt the very trees that were supposed to be protected.

From my own experience growing up in the foothills of Southern California, I am well aware of the danger of forest fires. We were evacuated many times in my youth. The home my family lived in at the time still stands—at least, as far as I know. But I've toured the area following some major forest fires in Southern California and many of the surrounding houses are gone. Now, living in the forest of New Mexico, those fire concerns still plague me. Following the devastating Los Alamos fires, there was a federal grant available to property owners to help thin the forest for fire prevention. Remembering the flames on the roof of our home in Southern California, I did not have to think twice before taking advantage of the opportunity. The forester explained that it was for the "health of the forest."

So, the spotted owl threatens private property rights, kills jobs, and puts the health of the forest in peril. But, at least we are saving the spotted owl, right? This is where the story gets really painful. None of this has helped to protect the spotted owl. The Oregonian states: "despite the economic sacrifice, the spotted owl population has continued to decline."

But it was never about the spotted owl. My New Mexico contact told me, "the spotted owl is the number one tool used by environmentalists to stop logging and shut down operations." Regarding this issue, Ron Arnold says: "In 1990, the forest issue was a hit for environmentalists in the Northwest. The northern spotted owl has become the celebrity bird for shutting down the timber economy under the Endangered Species Act." In an article titled, "Losing an owl but saving a forest," acclaimed author Jonathan Raban gives a description of the old-growth forests: "the squelchy, dank-smelling, multistory ecosystem, where the ground is strewn with rotting ancient tress, called 'snags' and 'nurse logs,' and the jungled undergrowth of ferns, vines, berries, moss and fungi..." It is about the forest that he says, "It was this irreplaceable ecosystem, centuries in the making, that environmentalists were really trying to protect under the terms of the act (whose first declared purpose is to conserve the ecosystem upon which endangered and threatened species depend); and the owl was their best available legal tool."

Okay. So, we've saved the forest. But wait, this was about the spotted owl. It was listed as endangered twenty years ago. If the experiment is a success, after twenty years, the owls should be multiplied to the point of being a pest. Instead they've continued to decline.

As it turns out, there are two major problems with this experiment. First, spotted owls do not need "old-growth forest." They have been spotted in younger, redwood forests where clear-cutting took place 100 years ago. They've been spotted there in greater density than in many old-growth forests. They have even been found in forests only thirty years old. And the birds are reproducing in these environments. Spotted owls have even been found raising their young in an artificial nest box.

My New Mexico contact told me stories of finding spotted owls in a variety of places including a crane bucket (Are you my mother? No, I am a crane) and on a derrick. Of course, no one wants to draw attention to the sightings, as the area will be shut down—killing jobs and hurting the economy in rural areas already hard hit by the lawsuits that block logging. These rural communities are plagued by a policy-induced poverty and do not want to invite more damaging government interference.

If the habitat of the spotted owl is obviously not that fragile and they can survive in old growth, new growth, a crane, or an artificial nesting box, what has caused their demise? Not humans.

The culprit is the barred owl who has moved into the neighborhood and is a more aggressive bully. Sometimes they kill the spotted owl. Sometimes they eat the spotted owl. Other times, they mate with the spotted owl creating what is called the sparred owl.

Wow! A whole new species. To me, a creationist, this sounds like the perfect evolutionary model. The gene pool of the weaker less adaptive species is diluted and a stronger version emerges. However, by definition, if they can mate, they are not two separate species.

But to those who believe that a set point in time is how things must stay, the natural progression of the owl is troubling. Environmental groups struggle with the ethics. Federal biologists have been experimenting with killing the aggressive barred owl in order to save the protected spotted owl. (Like the spotted owl doesn't know to stay within its "protected" boundaries, the barred owl doesn't seem to understand that the spotted owl has the full protection of the U.S. government.) Bob Sallinger, conservation director of the Portland Audubon Society, believes the highest priority needs to be placed on avoiding extinction for the spotted owl— which involves protecting the old-growth forest where the spotted owl lives. Regarding killing the barred owl, he said, "We are still struggling with where we come down." The fish and wildlife biologist said, "If we are going to remove them, a shotgun will probably be the method of choice because it is the most reliable."

Given the options for removing threats to the spotted owl, the timber industry should be happy that their death was by enviro-pressure, special interest group, and a kind of rural-cleansing—a slow death. But then again, maybe a shotgun would be more humane.

Ultimately, the only ones who've profited are Big Green and the lawyers.

Chapter 11

DDT

These days no one thinks much about DDT (Dichlorodiphenyltrichloroethane). Fortunately, for those of us in the Western world, today it is not something that requires thought. However, back in the days when the environmental movement was born, it was a hot topic popularized by Rachel Carson's book ***Silent Spring***, but has since faded from the memory of most Americans.

Its relevance here is that it was an early battle of the young environmental movement.

DDT was extensively used during the Second World War to control typhus and malaria. It was wonderfully successful. Millions of troops and refugees would have died had DDT not been used. Dr. Paul Muller, the scientist who discovered the insecticidal properties of DDT, was awarded the Nobel Peace Prize in 1948 for his work.

Following World War II, DDT was used extensively as an agricultural insecticide. Again it was successful. The U.S.

Department of Agriculture promoted its use because it increased crop yield by reducing the number of crop-eating pests.

This was a time of tremendous growth in America. More food was needed, and the crop increase was important—not just in America, but to help rebuild Europe. Consumables were in demand, and factories were running double shifts. DDT was just one of the many discoveries that helped create the time in America that so many look back to with great fondness.

DDT's use in the Western world was so effective that it did virtually eliminate malaria and other mosquito-borne diseases—so much so that, as NYT Editorial Board Member Tina Rosenberg said in an April 11, 2004 editorial: "We forgot why we once needed it."

While America was growing, waste was being dumped into rivers, smoke was pouring into the sky, and chemicals were being widely used—and perhaps, in some cases, misused. Eventually, our cities began to look like Beijing looks today.

It is from this backdrop—air and water becoming polluted and malaria a distant memory that Rachel Carson found an audience with her book ***Silent Spring***. One of the key premises of the book was that DDT was poisoning wildlife and the environment and endangering human health. The DDT scare is what was embraced throughout America's college campuses, gaining momentum, becoming one of the cries of the generation and filling a political void created by the end of the Vietnam War.

Carson cited experiments by Dr. James DeWitt claiming that DDT thinned the egg shells of birds, resulting in a reduced bird population—though Dewitt's 1956 article in *The Journal of Agriculture and Food Chemistry* showed that DDT-fed pheasants actually hatched about 50 percent more eggs than the "control pheasants."

In 1969, Dr. Joel Bitman, at the U.S. Department of Agriculture, fed Japanese quail a diet laced with DDT and found that they had thinner eggshells. His study, "DDT Induces a Decrease in Eggshell

Calcium," published in Nature on October 4, 1969, reported that the Japanese quail eggshells were about 10 percent thinner than those of quail who were not fed DDT—appearing to bolster Carson's claim. His findings were eventually overturned when, due to criticism, he redid the studies with sufficient amounts of calcium—the thinning disappeared. However, by then, the thinning eggshell theory was already part of the lexicon. Additional studies published in Poultry Science found chicken eggs almost completely unaffected by high dosages of DDT. But, those who championed the DDT ban based on Bitman's eggshell research wanted to hear nothing of the science. They had their cause.

In the 1970s, the debate was elevated and hearings were held before a Judge Edmund Sweeny, an EPA administrative law judge. Both sides of the DDT controversy presented testimony for seven months, equaling 9,000 pages of testimony. Ultimately, the judge determined that "The use of DDT under the regulations involved here do not have a deleterious effect on freshwater fish, estuarine organisms, wild birds or other wildlife." Yet the EPA banned DDT.

The science may have cause for some question about the widespread and careless use of DDT as was done in the '50s, but the use of political power is clear. The public pressure at the time of the DDT hearing, as a result of ***Silent Spring***, was so strong that the outcome was ignored in favor of the prevailing attitude.

At the same time, the environmental movement was picking up steam. The Environmental Defense Fund (as discussed in Chapter 8) was born in 1967. Banning DDT was its original purpose. Victor Yannacone was the cofounder and the original attorney for EDF. He tried the first action against the indiscriminate use of DDT and believes to this day that "The single lasting contribution of the Environmental Defense Fund to the protection of the world environment was its demonstration that the indiscriminate application of broad-spectrum persistent chemical biocides, such as DDT and the aptly named 'death group'—dieldrin, endrin, aldrin, toxaphene, and heptachlor—without regard to the phenomena of biomagnification and target species resistance was responsible for serious, permanent, and irreparable damage to the biosphere." While their Suffolk County, NY battle failed, they won the war.

William Ruckelshous was the first administrator of the EPA. Despite his earlier statement in a federal court about DDT's safety ("DDT has an amazing and exemplary record of safe use, does not cause a toxic response in man or other animals, and is not harmful. Carcinogenic claims regarding DDT are unproven speculation."), Judge Edmund Sweeny's exculpatory ruling, and the fact that he never attended the seven months of hearings or read any of the transcripts, in 1972, he banned the use of DDT. Ruckelshaus admitted that the decision was "political."

By then, malaria was no longer a concern for America or Europe, and no one balked at the DDT ban. Such is not the case for undeveloped countries that have seen an increase in malaria deaths since the DDT ban. Much of the funding for the DDT programs in Africa came from the developed world. As a result, several tropical countries had nearly wiped out malaria. Once DDT was banned, the programs were no longer supported and were shut down. The countries threatened by malaria did not have the funding or the infrastructure to continue them on a widespread basis, and malaria became Africa's largest killer. Since DDT was banned in 1972, billions have been stricken by malaria and tens of millions have died.

It is an interesting side note that many of the very same environmental groups who oppose the use of DDT are the same groups who worry about over-population. In a 1990 biographical essay, Alexander King, founder of the Club of Rome (neo-Malthusian movement) wrote, "My chief quarrel with DDT, in hindsight, is that it has greatly added to the population problem."

In 1998, the World Health Organization's Malaria Expert Committee held a meeting from which the following position emerged: "It is anticipated that for some time to come there will continue to be a role for DDT in combating malaria, particularly in the poorest endemic countries. Restrictions on DDT for public-health use contained in a future Persistent Organic Pollutants Convention should therefore be accompanied by technical and financial mechanisms to ensure that effective malaria control is maintained, to at least the same level, through vector control methods that depend less on pesticides generally, and on DDT in particular." On September 15, 2006, the World Health Organization announced that DDT "will once again play a major role in its efforts to fight the disease." They have given it a clean bill of health.

DDT has traveled a long journey since Dr. Muller was given a Nobel Prize in Physiology in 1948, to being banned in 1972 to being promoted in 2006. Yes, admittedly, the use being touted today is different from the widespread spraying being used in the 1950s. What is being suggested today, but still not widely accepted, is what is called Indoor Residual Spraying (IRS). IRS calls for the spraying of DDT on the inside walls of a dwelling and is especially effective on the kinds of walls found in the homes of the areas where malaria is the biggest killer: mud, concrete block, and raw plaster. The IRS application of DDT has a "triple-action" feature: it keeps mosquitoes from entering for six months or more, irritates the few who do enter so that they do not bite, and kills those that attempt to land. This use of DDT cuts malaria rates by 70 percent and in combination with other interventions, it can help reduce malaria by up to 95 percent. The DDT is being sprayed by trained personnel in selected locations, and it cannot be purchased in a local market, like we buy a can of Raid.

Regarding the IRS use of DDT, the World Health Organization—who used to promote the use of IRS DDT until the 1980s, when political pressure forced it to stop—now (2006) says: "We must take a position based on the science and the data. One of the best tools we have against malaria is indoor residual house spraying. Of the dozen insecticides WHO has approved as safe for house spraying, the most effective is DDT." (Dr. Arata Kochi, director of the WHO Global Malaria Programme.)

Even Victor Yannacone, the co-founder of the Environmental Defense Fund, has changed his position. He states, "I certainly do not oppose, and as presented, support, the limited application of DDT to the interior walls of houses in selected locations by trained personnel."

Despite all this, many groups such as Physicians for Social Responsibility and Pesticide Action Network still oppose DDT, even in the IRS use. The result is that these anti-pesticide zealots are perpetuating poverty, misery, disease, and death in Africa, Asia, and Latin America—a policy-induced poverty. Ugandan resident and malaria survivor Fiona Kobusingye-Boynes says that malaria "enslaves our continent, prevents people from working, keeps tourists and investors from coming to Africa and reduces the continent's economic productivity by billions of dollars a year." She calls the efforts of the greens, "eco-colonialism."

Bringing the topic up-to-date (as of the time of this writing), a group called Malaria No More is running television spots to draw attention to malaria deaths and raise money for campaigns. However, when you study their website, which says: "Malaria No More leverages high-impact awareness campaigns to engage the world, global advocacy to rally leadership and strategic investments in Africa to accelerate progress, build capacity and save lives" you find that IRS DDT use is not something they fund. With a goal of ending malaria by 2015, they do not fund IRS DDT use.

The science that caused DDT to be banned is spotty at best, the decision to ban DDT was admittedly political and many who originally opposed DDT now acknowledge that it is the most powerful, effective, long-lasting mosquito repellant ever invented and has been proven to virtually eliminate malaria—yet environmental groups still oppose its use.

How many people have to die for enviro-radicalism before we quit listening to them altogether?

Chapter 12

Ethanol

A friend calls you up. "I need some advice. Can we meet for a beer?" "Sure," you say and set a time and a place.

"It's my son. He is thirty years old. He is still living at home. I have to give him more and more money. When he was young, I expected to pay for all his expenses. I mean he was just a child. But now, he should be on his own. He keeps telling me he is going to get a job, but you know it is a tough market out there.

"In fact it is so tough, my pay has been cut. I have to trim all my expenses. But he doesn't understand. He claims that he still needs the same amount of money. I don't know how long I can keep this up. I have to borrow to support his habits.

"His girlfriend used to like that he didn't work. He had lots of time, and my money, for her. But even she has tired of his tricks. Even she is saying that the relationship isn't working. That it was a mistake in the first place.

"What should I do?"

You laugh at your friend. "You fool. You are too *nice*. You are being taken for a ride! You need to boot your son out. Until you make him stand on his own, he is going to keep asking for more and more."

Any reasonable person would offer the same advice.

In this analogy, the son is ethanol. It has been supported in America for thirty years. It is reasonable to expect to help the new industry get off the ground, but, if after all this time, it still needs dad's help, it is time to boot it out.

The father is the United States Government who keeps throwing good money after bad. At a time when everyone else is cutting back, and all plans need to be trimmed, the son—ethanol—is still getting the same subsidy. Dad, the government, is borrowing money from China in order to keep the ne'er-do-well son funded.

The girlfriend is Al Gore. He used to heap praise on the son—ethanol, but now, even he has seen through the ruse. Even Al Gore acknowledges that ethanol was a mistake.

Ethanol actually has a long history in America. It was the fuel for the early automobiles. But then oil was refined into gasoline. It was much more effective than ethanol and, in the free market, ethanol producers went broke.

It made a re-appearance in the 1970s, in the days of gas lines and high prices. Americans were in a panic; we had to do something to loosen the stranglehold the Middle East had on us. Ethanol, then called gasohol, was touted as the answer.

It sounded good. We can grow it, and we can keep growing it. It is "renewable." Lobby groups were formed to promote government funds pouring into the farm states. It didn't hurt that one of the key corn-growing states is Iowa—the same state that does the initial picking of presidents. No one with aspirations for the White House can afford to go against helping Iowa.

Ethanol did nothing to cut our use of Middle Eastern oil, but it still didn't go away.

The modern ethanol renaissance began in 1990 with amendments to the Clean Air Act.[96] It regulated not just how to burn gasoline, but how we make it as well.

Its next big boost came due to a new panic: global warming. We were told that the earth was warming, the ice caps were melting and that polar bears were going to die. Al Gore's movie, *An Inconvenient Truth*, led us to believe that cities would be underwater if we did not curb our fossil fuel use. Add that to the fact that relations with the Middle East have gone downhill and ethanol is back in the news. It became America's energy hope. President George W. Bush poured money into it.

To support this fledgling industry Congress passed a variety of incentives to help it succeed.

- Ethanol gets subsidies—ultimately paid for by the taxpayer. The Volumetric Ethanol Excise Tax Credit, a 45-cent-a-gallon subsidy for ethanol blenders, was passed by Congress in 2004. U.S. blending tax breaks for ethanol make it profitable for refiners to use the fuel even when it is more expensive than gasoline.
- Ethanol is mandated. In 2007, motor-fuel retailers were required to blend at least 15 billion gallons of ethanol per year into the nation's gasoline supply by 2015.
- Ethanol is protected. While ethanol can be made more cheaply from sugar in Brazil, we do not want Brazil's ethanol. To be sure that the young American industry can survive, imported ethanol is stuck with a 54-cent tariff—making it noncompetitive with the home-grown, corn version.

In a 2010 letter to Senate leaders,[97] seventeen senators from both sides of the aisle urged them to end ethanol protections. They pointed out that the current subsidy was "fiscally indefensible." The letter states: "The data overwhelmingly demonstrates that the costs of the current ethanol subsidy and tariff outweigh the benefits." The senators point out that ethanol is the only product to receive a trifecta of government support: direct subsidies, mandates that force consumers to buy ethanol, plus trade protectionism—typically support comes in the form of just one. Despite this logic, the ethanol subsidies continue.

But it is green and renewable. The burning of ethanol as a fuel doesn't emit CO_2 into the atmosphere. So it is worth the investment, right? Not according to the King of Green, Al Gore. Though he supported huge subsidies for ethanol in the 1990s, Gore has now acknowledged that it was about what was best for his presidential run, not what was best for the environment. In a speech in front of a green energy conference in Greece, Gore said, "First generation ethanol I think was a mistake. The energy conversion rates are at best very small."[98] Additionally, he admitted, that the push for corn-based ethanol did raise food prices.[99]

While exact numbers vary from study to study, even the most pro-ethanol admit that it takes about a gallon of petroleum to produce a gallon of ethanol. The ratios are often found to be more like three to one. Studies have concluded that ethanol increases greenhouse gases (GHG) and is a net energy loser. The Congressional Budget Office found

> When Al Gore drops an environmental fad, it has truly reached its expiration date.[100]
>
> Rich Lowery, Editor, *National Review*

that using ethanol to cut GHG cost about $759 per ton when the market value of a ton of GHG, according to the European Climate Exchange, is about $19. Greens are abandoning ethanol—but not Congress.

Ethanol has other flaws too. It contains less energy than gasoline, which means that if your car typically gets twenty-five miles per gallon with gasoline, with E10—a 10 percent ethanol blend—you are apt to get only 21 mpg. With E85, your mpg could be as low as 16 mpg. At the same time government has mandated that that mpg performance increase by 57.7 percent, it has also mandated increased use of ethanol.

Not only does ethanol wreck your mpg, it wrecks your engine. Ethanol can corrode engine parts and damage catalytic converters. Some newer vehicle manuals warn that use of E15 could void the warranty.

At Abe's K & S Services Center, they specialize in small engine service and repair. Abe says 85 percent of the repairs they do are caused by fuel problems. Because of the increased ethanol in the fuel available at gas stations, Abe's had to change its warranty policy. The center no longer warranties fuel-related damages. He steers people to pure-gas.org, which shows where customers can buy ethanol-free gas nationwide. For his customers, many of whom are in the lawn-care business, the increased ethanol doubled their repair costs until they learned about its hazards and quit using it. The ethanol eats up the rubber components like a cancer. It has no shelf life. Someone with a weed whacker that is used sporadically may find that, after sitting for a few months, the unit won't start. Abe looks at it and fixes the rotted fuel lines. The customer leaves happy. They whack weeds and put it away. A few months later, they take it out for a day of yard work and find the same problem. The ethanol has broken down and washes away lubricants. For customers who use the small engines infrequently or those who depend on the equipment for their livelihood, Abe recommends TruFuel, an ethanol-free, premixed fuel for small engines that has a shelf life of up to five years and costs about $6.00 a gallon.

In Albuquerque, city vehicles have gone back to regular unleaded gas rather than the more expensive flex fuel they'd been using for about four years.[101] Fire Department Chief Mark Garcia addressed the concerns: "We do have problems when we are talking maintenance. We want to make sure the fifty-five emergency vehicles we have are ready at the spur of the moment." Police Department Chief Ray Schultz had similar concerns, "When we first switched to E85 we were seeing dozens of vehicles because of the corrosive nature of the fuel. It was mainly a fuel pump issue." The city is saving money by using less expensive regular unleaded gas, and they are saving money with fewer repairs.

Ethanol isn't just bad for your car and your budget—it damages wildlife and fish. In the April 2007 issue of *Fly Rod and Reel Magazine,*[102] America's foremost conservation writer, Ted Williams says, "Instead of cleaning up America, ethanol has added to the mess we are making out of our water and air." He asks, "How will ethanol affect your fishing, apart from possibly ruining your outboard motor?" Here's his answer: "First, no crop grown in the United States consumes and pollutes more water than corn. No method of agriculture uses more insecticides, more herbicides, more nitrogen fertilizer. Needed for the production of one gallon of ethanol are 1,700 gallons of water, mostly in the form of irrigation taken from streams either directly or by snatching the water table out from underneath them. And each gallon of ethanol produces twelve gallons of sewage-like effluent." What does this "effluent" do? Williams says, "The toxic, oxygen-swilling stew of nitrates, chemical poisons and dirt excreted from the corn monocultures of our Midwest pollutes the Mississippi River and its tributaries, limiting fish all the way to the Gulf where it creates a bacteria-infested, algae-clogged, anaerobic 'Dead Zone' lethal to fish, crustaceans, mollusks and virtually all gill breathers."

No wonder Al Gore has decided it was a "mistake." But Congress continues to support it. Not because it is a "better mousetrap," not because the science supports it, not because it costs less, or makes life better. History is already showing that ethanol, is one more green scheme. It doesn't perform as advertised, but it costs more.

The promotion of ethanol is one more example of a government out of touch—cutting what we are good at and have historically built on, to fund what doesn't work and the public doesn't want. There is only so much money (which is now gone) to spend on stuff that doesn't work and doing so is irresponsible—especially when there are vital projects going unfunded.

Wind energy is virtually thc same, but its early supporters have not turned on it yet. And, it does not have the trifecta of government support—though if there could be a tariff on imported wind, there would be. Neither is based on sound science. Both are the price of *nice*

It is time to boot the ne'er-do-wells out. We can't support the indigent son any longer.

Section Four

Act Now!

Chapter 13

Wake Up, Show up, Stand Up, and Speak Up

"What do you do?" asked Leroy, a forty-seven-year-old welder doing some work at my house.

Hmmm, how do I describe what I do? It is not like I am a nurse, a teacher or a lawyer. I've attended the same church for five years and most of the people there know me as the "coffee lady" because I run the espresso bar on Sundays. They are surprised when they read an op-ed that I had written published in the paper, hear me on talk radio, or see me on television. They are learning I do more than make a mean espresso.

I gave Leroy my standard answer: I run a non-profit organization that advocates on behalf of energy and the energy industry. Some people, like those sitting next to me on airplanes, nod and change the topic. But Leroy persisted. He and I had a conversation standing in my front yard—much like those I have with individuals who come to talk to me after I speak.

Leroy said, "Most people in my generation had confidence in government. We never thought to question. We were busy living our lives. Now, we've awakened. We see our freedoms slipping away."

Following his comment, I launched into a response. He sat on my stone wall while I stood like a teacher with a lone student.

Like Leroy's generation, you've woken up. Hopefully the previous chapters have been a wake-up call to you. Probably like Leroy's generation, you'd already been alerted as you've seen your freedoms slip-sliding away. This concern may have prompted your interest in this book.

In Chapter 7, Smash the Watermelons, we looked at who the "watermelons" are and their history. We learned how they got to where they are, but we did not address why we let them take over—Leroy's comment.

So what is our role?

There have probably always been people who do not like America—but they became more outspoken in the 1960s. They voiced their opinions in protests and rallies. They seemed extreme. Most of America ignored them.

We went to work, raised our families and lived our lives. While most of America was busy with what went on inside our four walls, there have been people who have been planning and plotting to change America. No, I am not talking a conspiracy theory here—though there may have been one (still may be one). They have been subtly working behind the scenes and most of us have paid no attention. They have been pushing change via incrementalism. We did not notice. We liked our lives. We were happy with America.

Now those same people are in Washington. They are in charge. When they had control of all three houses—the White House, the Senate, the House of Representatives—the incrementalism went out the window, and they went full-bore. We were sleeping, and our freedoms eroded. Then, we woke up.

This is where most Americans are now. Awake. Like someone who's heard a noise in the night, we are awake, but scared—pulling the covers up around us.

If the noise in the night were really something of concern, sitting in bed surrounded by comfort, would not chase away the scare. We'd need to scream to let them know we are there, call 911, and grab a weapon.

With the scare we are feeling in America, we need to do more than wake up. We need to act. We need to show up, stand up, and speak up.

You may feel overwhelmed—thinking: "What could one person do?" Many feel the die is cast; there is nothing we can do. This is where the show up, stand up, and speak up theme comes in. Waking up is not enough. We need to take action—and we can make a difference.

Remember cap and trade? This was one of President Obama's campaign promises. He declared that he would pass cap and trade. With those in support of the carbon emission limiting legislation in control of all three houses, it should have passed easily.

Known as the Waxman-Markey Climate Change Bill, it did pass the House six months after President Obama took office, June 26, 2009.[103] It was expected that the Senate would follow suit before they went home for the August recess. But the Senate had a full schedule and cap and trade did not make it to the Senate floor for a vote before they left Washington on August 8. Once back in their states, Senators and Representatives faced their constituents—and they weren't happy. Remember town halls and Tea Parties? Those who voted for Waxman-Markey in the House got an earful. Senators learned that the public did not want cap and trade. The public made a lot of noise, the Senators listened. Cap and trade never passed the Senate. It never became law.

Don't ever doubt that "we the people" have power. But, we must show up, stand up, and speak up—just like we did in the summer of 2009.

Built upon the sand

(24 September 2009)

Now that summer recess is over for Congress and our elected officials are in D.C., cap and trade is back in the news. Many people are asking, "Just what is cap and trade?" Last week while I was speaking in Canada, a woman questioned what she'd heard. At church, a man asked me about it. And, my best friend Patti brought it up while we were getting a pedicure. There is vast misunderstanding as to what "cap and trade" really means.

Cap and trade reminds me of the old song from children's church, "...The wise man built his house upon the rock...The foolish man built his house upon the sand...the rains came down...the house on the rock stood firm but the house on the sand fell flat." Cap and trade is like the house built on the sand. It does not have a firm foundation.

First, the cap-and-trade house is built upon a lie. The lie is that climate change is a crisis that it is caused by man, that man can fix it, and that even if no other countries participate in "capping emissions," America can still make a difference.

But what if it is not a lie? What if the crisis is as real as the global cooling crisis was perceived to be in the 1970s? If that is the case, the crisis can wait a few more years. We've been making noise about climate change for forty years. Now, in the midst of a major economic recession, is not the time to be enacting extreme measures that will finish off the American economy.

If it is a crisis, and CO_2 is responsible, a carbon tax is more honest—people will know where the price increases come from and blame the government. With cap and trade, the high costs are initially borne by energy producers and manufacturers who will pass costs on to consumers. This way, anger over price increases will be deflected to the big, bad companies and government will appear virtuous.

Additionally, cap and trade will create a whole new level of bureaucracy—and we all know how effective layers and layers of bureaucracy are. Those who are able to cut CO_2 emissions will get credits. Those who emit will have to buy the credits from those who saved. The credits will be traded like stocks on Wall Street—hence "trade."

While cap-and-trade proponents, such as Nancy Pelosi, tout the availability of "green jobs" due to the passage of the Waxman Markey Bill, this is "happy talk" as *The Economist* Magazine called it. Some short-term installation jobs may be created, but the bulk of jobs generated due to the government mandate for wind and solar energy will be overseas. While the turbines turning or the solar panels collecting sunbeams do not produce CO_2, like a cost-effective coal-fired power plant, manufacturing the turbines and solar panels does—which is why much of the production is done overseas where the regulations are more conducive to industry. Here are but a few examples of foreign involvement in America's so-called "green energy:" Mitsubishi's 90-foot- and 100-foot long turbine blades are manufactured in Mexico. Nine of the top ten wind turbine manufacturers are NOT in America. Each wind turbine contains two tons of rare earth—provided by China.

This so-called "green energy" is expensive, while the carbon-based energy (oil, natural gas, and coal) is cheap by comparison. Cap and trade, the Waxman Markey Bill, will increase the cost of energy by as much as $1,700-$3,000 per household—and the latest news reports indicate that the Obama administration has known this all along. Energy producers will have to up the prices to consumers. Suddenly "green energy" will have a competitive chance—but the average household will not be able to afford it. Proponents claim that tax credits for the poor will offset the higher costs. But, where does that leave most of us?

To anyone without a clear concept of "cap and trade," the simple way to explain it is like the old Papal indulgences that allowed you to sin, if you paid. In this new world, CO_2 is sin. Those who do not sin will be rewarded and the "sinners" will have to pay. The Pope (some government bureaucrat) will be able to determine who gets rewards and who has to pay.

Hardly the energy freedom we've become accustomed to in America. Cap and trade is much more like energy socialism.

Because of the public outcry exhibited in 2009, many elected officials no longer hold the large town hall-style meetings. The shocking shooting of Representative Gabrielle Giffords in January of 2011 has given additional cover to those who do not want to face angry crowds. The good news is that, while your representative or senator may not want to face you, you can still make your voice heard if you show up, stand up, speak up—even if only figuratively.

There are dozens of hearings and meetings each week that include city, regional, state, and national issues. They need you! These are usually held during normal business hours and have minimal mention in the media—meaning that you have to be alert to the issues and track what is going on. They often slip by unnoticed until the plan they implemented is enacted and people realize another freedom has disappeared. The timing of the forum for public comment makes the participation of citizens who are retired, students, or self-employed imperative. These are also a wonderful learning opportunity for home-school groups. The parents can speak up and the students can observe and learn about their government.

Back in February 2008, I cut my teeth on this type of meeting at a hearing for a proposed Traditional Cultural Property (TCP) designation for Mount Taylor in my home state of New Mexico. I received a call from one of my contacts after business hours on a Thursday evening. The designation would greatly hinder uranium mining that was ramping back up in the area known as the "uranium belt" in the rural, economically devastated region near the town of Grants. Due to the late notice of the meeting, no one but a few lawyers from the industry could attend. They asked me to go. I did. The experience was eye-opening. TCP supporters, people who opposed the uranium-mining renaissance were there en masse. They had prepared scripts and spoke in favor of the TCP one after the other. Representing the mining interests, normally articulate attorneys scratched down some notes and then begged for a two-week extension. At the end, I stood up and spoke, too. I did not know much about the situation, but I could not keep quiet.

A few minutes before 5 p.m., the board deciding on the TCP took a recess. After a brief discussion, they came back, voted for the TCP, banged the gavel and dismissed. Supporters cheered. Opponents were shocked. I woke up—literally.

Renewed mining can help financial woes
(30 September 2009)

The energy industry's presence once gave New Mexico the luxury of a budget surplus. However, tighter and tighter regulations are squeezing producers and many are leaving the state or choosing not to do business here. The oil and gas industry will no longer be a cash cow. Fortunately for us, nuclear energy is in a global renaissance and it needs uranium—which New Mexico has in abundance.

We complain about the high cost of gasoline and heating. However, unless more base-load energy sources are developed, our electricity costs will continue to climb as energy becomes scarcer and more costly.

Last week, something could have been done to help New Mexico's energy picture. Instead, the Department of Cultural Affairs Historic Preservation Division stacked more damaging regulations on to the policies already hurting the state's economy.

Regulations are already in place to ensure that the environmental impact of extractive industries is minimal. Industry advancements give mining a smaller footprint and safer practices than fifty years ago. Archaeological-impact studies provide protection. Executive orders require communication between involved parties. Miles of red tape are already involved before any drilling or mining can be done.

Friday afternoon, a hearing was held to determine whether or not to give Mount Taylor an emergency listing in the State Register of Cultural Properties. Unfortunately, announcements about this important meeting were distributed a mere two days earlier. That's not fair.

The attorney who'd prepared the nomination gave an impassioned—and well-prepared—plea to the Cultural Properties Review Committee. A line-up of supporters was prepped with written presentations. They gave emotional explanations as to why this 600-square-mile region should be awarded the Temporary Cultural Property listing. The whole thing was rigged.

No one who might oppose the nomination of Mount Taylor as a TCP had time to prepare his or her side of the case. Potential opponents were given time at the hearing to present, but none of us had ever seen the nomination or the supporting maps and legal descriptions. Not fair!

Three different attorneys stood up and begged for a postponement of two weeks to review the documents. Star Gonzales, Executive Director of the Grants Chamber of Commerce, urged the committee to hold another hearing in two weeks "in the shadow of the mountain," in Grants, so stakeholders could participate. Heirs to the impacted land grants requested time to review the proposal. After watching all possible opponents hurriedly scribble out some notes in hopes of presenting an articulate argument, I, too, petitioned for an extension.

The Nuclear Regulatory Commission granted a six-week extension to Tribes and environmentalists on a different issue in August. Shouldn't a State Agency have the same flexibility and courtesy?

"What is the 'emergency'?" Uranium mining exploration permits have been issued for the Mount Taylor region.

> "What will this nomination do?" Reportedly, give the impacted Tribes communication about the mines. An executive order signed by Governor Richardson already requiring this was brought up and dismissed as irrelevant.
>
> "How will this impact my rights to the land?" A land grant heir to land impacted by the nomination, repeatedly asked. Katherine Slick, Director of the Historic Preservation Division, finally admitted that the TCP listing would add "red tape" or "layers of bureaucracy" to any new permit requests—though she preferred to call it a much softer term: "process."
>
> After nearly three hours of entreaty and theater, just before 5 p.m., they granted the TCP. Just like that, it was done. It was clear that the committee had already made up its mind and was only holding the hearing as a formality. The voice of the local people who have lived in the shadow of Mount Taylor since the 1700s, farming, ranching, building homes and communities and—yes—mining, fell on deaf ears. Cultural Properties Review Committee: shame on you! It's time we make these insiders explain their overwhelming influence to a judge.

The next morning, I woke up with one of the first op-eds I'd ever written on my mind. My furor had percolated all night. I got out of bed, went to my computer and pounded out "Renewed Mining Can Help Financial Woes."

Multi-year story made short (2008-2011): the meeting was declared to be in violation of the New Mexico Open Meetings Act. The next year was filled with hearings and public meetings. Overflow crowds showed up, stood up and, spoke up. As a result the issue got media coverage it would have never received without the crowds. Ultimately the committee made the same decision they did in that sparsely-attended meeting on a snowy February afternoon in 2008. The committee chair was rewarded with a position in Washington. Since then a lawsuit ensued, and the entire thing was reversed by a state district court.

Had the public not gotten engaged, had the issue received no media coverage, it is likely that the original decision would have held.

A similar story happened with New Mexico's statewide cap-and-trade proposal. The difference is that the Environmental

Improvement Board hearings took place in 2010—after the national public "wake up" occurred. The hearings drew crowds and public comment was offered from both sides.

Since the Mount Taylor issue, we've learned. I sent out notices about the meetings, as did the state's free market think tank. Back in 2008, there was not the network that exists today. Our announcements this time were picked up and sent out via e-mail to myriad other groups (groups, which did not exist in 2008).

One retired couple served as stalwarts for the issue and attended virtually every meeting over a ten-month period. They wrote up reviews of the meetings and sent them out to an interested e-mail list. The joint efforts resulted in a strong showing opposed to the New Mexico-only cap-and-trade program.

Both of these cases were about energy—hence my involvement. I've participated in other energy-related hearings and meetings since my first in 2008. Some have been before appointed committees and boards, others with the state Legislature.

Regardless of the setting, I have learned that the energy killers, the Watermelons, Big Green, have been at this much longer than all of us who want abundant, available and affordable energy. They are plugged into networks, and they show up at the meetings. They stand up and speak up. Without a balancing voice, the decision-makers almost have to believe that the group who showed up represents the voice of the people. We've been silent, while they have been speaking up.

Your voice is needed. Your voice matters. Not just for the big rallies and town hall meetings, not just for the hearings, but through phone calls and e-mails. This is another way you can show up, stand up, and speak up—and you do not even have to go anywhere or carry a sign.

Committee and board members and elected officials at every level may want to vote against the energy-killing policies, but they need cover. They need to be able to quote volumes of responses they've had on the matter. Those responses can be in person, via e-mail, or by phone. Elected officials can often be heard on the news stating that their offices have been overwhelmed with calls and e-mails over a particular bill. They cite the numbers as an indication of how strongly their constituency feels about how they vote.

As I have spoken on energy to citizen, activist, and industry groups throughout the country, I often hear frustrated comments like, "All they care about is getting re-elected." I wag my finger back, pointed at his or her face, and say, "And therein lies your power." Yes, they do want to get re-elected. But if they repeatedly go against the will of the people, they will not win their next campaign. Elected officials often break with their party, their voting record, and even with their former campaign promises when faced with a tough election. Some call this flip-flopping. On the positive side, they may well be responding to political pressure—pressure we can apply.

The easiest way to speak up is to have your representatives' phone numbers programmed into your phone. When you have an unexpected wait, simply dial their number and tell them how you feel on an issue. Praise for a good vote is helpful, too. Let them know they are being watched.

E-mails and faxes are also helpful. Don't bother sending a paper letter. Due to security, paper doesn't actually get to Washington D.C. offices. Letters are often delayed by off-site digitizing and are then delivered electronically.

Routed greens retreat (14 April 2011)

Climate change is real. Climate change is manmade. Manmade climate change has happened within the past twenty-four months.

Leaders in the climate change debate have controlled the message for forty years since the adoption of the Clean Air Act. They have "approached climate change politics with an air of disdain," according to Fred Krupp, President, Environmental Defense Fund (established in 1967).

Krupp addressed the changing political climate at *Fortune* Magazine's Brainstorm Green Conference in early April and admitted that there is a "newfound hostility to climate policy." He advised the environmental community to be "more humble" and "less arrogant." He acknowledged the failure of a comprehensive energy and/or cap-and-trade policy.

Krupp is correct. With the falsification of climate records exposed—known as Climategate—the American people now see climate change as merely hysteria. Polls show they do not view it as a real problem that we need to address now.

At the same conference, Jim Rogers, CEO of Duke Energy, agreed. He said, "Cap and trade cannot be sold and must be reinvented," adding that it was going to be hard to "resurrect cap and trade."

Climate change legislation has been the holy grail of the environmental movement—but the climate has changed. Now the green movement is playing defense.

This change of climate is not from carbon emissions—though it is manmade. At the same conference, former green jobs czar, Van Jones, didn't "blame the environmentalists or the policy concept itself." What brought about the change? How'd the debate get reframed and cause the death of cap and trade? Something, Jones pointed out, no one in the room had heard of twenty-four months before— "a right-wing populace movement" that the carbon emission supporters failed to take seriously: "the tea party."

Man changed the political climate in just two years. After forty years, environmentalists are now on the defense because of some underestimated "upstarts."

This change of climate was evident during the recent legislative battles over the Environmental Protection Agency's (EPA) authority to use the Clean Air Act to regulate carbon emissions—which environmentalists believe causes climate change.

In late March/early April, four bills were introduced in the Senate—each designed to limit the EPA's authority. Not surprisingly, none passed in the Democrat-held Senate. However, the change of climate can be seen in the numbers. In the vote on April 6, the bills with the least restrictions on the EPA's authority were trounced (Baucus Amendment 7-yes, 93-no; Rockefeller 12-yes, 88-no; Stabenow 7-yes, 93-no) while the strongest, the McConnell Amendment, came close to passing at 50-yes, 50-no. A similar plan passed the House 255 to 172.

While the climate has changed, the fight is not over. The battle continues. Following the Senate's EPA skirmish, *The New York Times* said: "Efforts to handcuff the environmental agency are not likely to end here."

Admitting defeat on climate change legislation, Krupp encouraged the environmental movement to try "other approaches."

As soon as Krupp's suggestion surfaced, "other approaches" appeared. The next day (April 6) Senators Tom and Mark Udall (NM and CO, respectively) introduced legislation that retiring Senator Bingaman (NM) has been heralding for several years: The Renewable Energy Standard (RES). Simmering on the sidelines, the RES was ready and awaiting its moment. With the admitted death of cap and trade and animosity toward the EPA growing, it was time for something, as Krupp stated, "that might capture a bipartisan center." The cousins Udall were all too happy to oblige with a bill that would set a federal RES of 6 percent renewable energy (wind, solar and "other renewable sources") by 2013, reaching 25 percent by 2025. Both New Mexico and Colorado currently have state renewable energy standards—as do twenty-seven others.

The Udalls' efforts, in this changed political climate, have so little chance of success major news outlets ignored their announcement. *The Colorado Independent* said: "While the bill may be able to make it out of the Senate—although even that isn't a certainty—it has almost no chance in the Republican-controlled House."

The RES should be seen as the "reinvention" of cap and trade—another approach. According to *The New Mexico Independent*, "The plan would allow energy suppliers to buy credits from other producers who produce renewable energy and allow producers to 'bank' the credits for up to four years and borrow credits for up to three years into the future."

The RES would essentially achieve the same carbon emission reductions as a cap-and-trade plan by forcing the public to use more expensive renewable energy—thus reducing energy consumption. (Generally states with a renewable energy standard have higher electricity rates.) Americans understand that when developing countries refuse to cut their energy use because it will hurt their economy, we shouldn't be forced to cut ours either.

Despite the near certain failure of the Udalls' approach, don't make the same mistake the White House made. Take these repeated "resurrections" seriously by keeping the green movement on the defense. They've been trying to make us replace economic energy that works for electricity that is expensive, intermittent, and ineffective. But America noticed. We woke up, showed up, stood up, and spoke up.

You have changed the climate!

Regardless of how you contact your congressional representatives, don't be intimidated or think you do not know enough to call or write. The harsh reality is the elected officials—

especially state or national office holders—are never going to see your letter. A staffer will simply note which side of the issue you represent and may send you a generic response. I've had people who I've motivated to write their senator or representative on an issue express frustration over receiving a letter back that totally misses the point of what their letter said. They spent time researching their opinion and providing a thoughtful argument, only to get a response as if they'd supported the very thing they oppose.

Just make the call or send the e-mail asking them to vote for or against the issue at hand.

Which brings us to the next question, how do you maintain your real life and stay abreast of when you should go to a hearing or write/call an elected official? You communicate with groups or organizations that support your views. Before an important vote or hearing, we send out e-mail alerts asking for your participation. Most groups are networked with others and we send out announcements for one another. Begin to build an e-mail list of your friends and then forward alerts on to them.

We focus on energy issues statewide, region-wide and nationwide. Energy Makes America Great Inc. works closely with several other groups. One is the state free-market think tank. They deal with taxes, education, and other statewide issues. Most supporters of abundant, affordable, and available energy are also free-market supporters. If I receive an announcement from them on an education issue, I do not pass it on—even though I may personally agree with it. That is not my focus. But we did work together on the statewide cap-and-trade proposal—it was a state issue, a free-market issue, and an energy issue. Different groups are needed to watch specific interests because there is too much happening, too quickly, for any one entity to track them all.

Within energy, there are groups such as the American Petroleum Institute, the American Coal Council, and the Nuclear

Energy Institute. Each deals with a single aspect of energy. Then there are groups like the Competitive Enterprise Institute, the National Taxpayers Union, Americans for Prosperity, Committee for a Constructive Tomorrow, and the Heartland Institute, which address a broader range of topics that sometimes includes energy. Energy issues also intersect with off-roaders and property-rights folks. At Energy Makes America Great, we interact with all of them and pass on alerts—which we usually repackage—when appropriate.

If you already have groups you trust—such as Energy Makes America Great Inc.—be sure you sign up for their alerts. Then, act on them!

Another source is talk shows and newspapers. When a viewpoint you support is presented, respond. Contact the talk show or write a "letter" to the editor (LTE). Realize these media are all about advertising dollars, and they cannot get advertisers without viewers/listeners/readers. They need to know what their audience wants; they need to keep you. Once again, the watermelons are already doing this. If the only people who write to the newspapers are energy killers, the paper assumes that is the viewpoint people want. These outlets also thrive on controversy—they like a healthy back-and-forth.

I spoke at a rally about a Forest Service road-closure plan. A standing-room-only crowd of 750 people attended. Following the meeting, a newspaper article misquoted me. A week later, the newspaper featured several letters to the editor against me and the Congressman, who was the featured rally speaker. But this is good. It brings attention to the issue. A few days after the letters were published, I spoke for a group in the same community. People said they'd come because my enemies were their enemies. Follow-up LTEs in support of the Congressman and me kept the issue in the public dialogue even longer.

If you write a letter to the editor and it doesn't get published—especially in a major metropolitan area—do not be discouraged. The bigger papers get more than they can publish, but like contacting your elected officials, your voice still gets counted. You've spoken up.

For talk shows, guest/topic success is gauged in response. I've been kept on talk shows for hours because the phone lines have been consistently lit up. Before I go on the air, we send an e-mail to our newsletter distribution list announcing the show and asking for callers. People often listen on the Internet and call from across the country—but the lines are lit up, the host is happy, and the message gets extended exposure.

Like the road-closure rally, showing up at public events is also important. The crowds are needed to get media attention. If only twenty-five people had attended the road closure rally, the television stations and newspapers would not have covered it. Because the event gets the "earned publicity," more people find out about the cause or issue. More media attention begets more media attention. And the impact can snowball.

Like my welder Leroy and his friends, a lot of people have awakened. But changing the course of public policy takes more than being aware of the problem. It takes action. Will you use these practical suggestions to show up, stand up, and speak up? I believe together we can turn back the tide. Instead of the "environmental utopia" we are headed toward, we can live comfortably and confidently in American exceptionalism—fueled by energy.

This book is about energy. But as an aware American, you know about the many issues facing our country today. I trust you have become awakened, activated, and energized to show up, stand up, and speak up on a variety of issues. Most importantly, I hope the connection between energy and freedom is clear to you. If they take away your energy, they have taken your Freedom.

For additional information, updates, and action alerts,

go to www.EnergyMakesAmeicaGreat.org.

Be sure to sign up to receive the FREE e-newsletter:

The Voice for Energy.

Appendix

Environmental Utopia

What Would Life Be Like If the Environmentalists Were in Charge?

An analysis conducted by
Citizens Alliance for Responsible Energy (CARE)
Marita Noon, Executive Director
John McCulloch and Lawrence J. Clark, Research Assistants

The following table contains a random sampling of environmentalist organizations with goals and/or current campaigns designed to interfere with, reduce, or eliminate energy-related activities in the United States and around the world.

<table>
<tr><th>Stated Goals</th><th>Actual Energy/Mining Related Projects</th></tr>
<tr><td>Caribou Clean Water Partnership (Idaho & Wyoming)</td><td></td></tr>
<tr><td><ul><li>Balance the private interests of the phosphate companies</li><li>Protect Caribou National Forest from selenium contamination due to irresponsible phosphate mining practices[104]</li></ul></td><td><ul><li>Convince BLM and Forest Service to deny Simplot's bid to obtain ownership of any BLM or Forest Service lands for their toxic tailings impoundment</li><li>Convince BLM and Forest Service not to approve any lease modifications for lands within the Huckleberry Basin roadless area[105]</li></ul></td></tr>
<tr><td>Center for Biological Diversity</td><td></td></tr>
<tr><td><ul><li>Fight a growing number of national and worldwide threats to biodiversity, from the overarching global problems of overpopulation and climate change to intensifying domestic sources of species endangerment</li><li>Secure broad protections for species and habitat[106]</li></ul></td><td><ul><li>With their campaigns to protect goshawks and owls, they shut down major timber operations throughout Arizona and New Mexico</li><li>Brought an end to large-scale industrial logging in the heritage public lands of the arid Southwest[108]</li><li>Aided in the removal of two power plants on Fossil Creek[109]</li></ul></td></tr>
</table>

• Make sure logging's carbon consequences are addressed • Stop dirty energy development • Preserve ancient desert tortoise • Keep off-road vehicles from tearing up delicate ecosystems and imperiling species throughout the United States[107]	
Clark Fork Coalition (Montana)	
• Dedicated to protecting and restoring the Clark Fork River basin[110]	• Opposes ExxonMobil's proposed "high & wide" corridor in Montana[111] • Stop or re-route mines, pipelines and other riverside growth • Helped outlaw cyanide heap mining • Aided in stopping Rock Creek Mine from opening • In May 2010, Federal Judge Molloy rejected the U.S. Forest Service's approval of a proposed silver and copper mine (known as the "Rock

	Creek Mine") on the edge of the Cabinet Mountains Wilderness Area within the Kootenai National Forest. Judge Donald Molloy ruled that the Forest Service violated the National Environmental Policy Act and the Forest Service Organic Act in approving the Rock Creek Mine planned by Revett Minerals Inc.[112] • Appealed permit for Thompson River Power Plant[113]
Coalition for Otero Mesa (New Mexico)	
• Protect Otero Mesa from risk of ruin from exploration for oil and gas deposits • Prevent drilling in the Otero Mesa Wilderness[114]	• Mobilized a coalition of hunters, ranchers, conservationists, and state leaders to stop any new oil and gas drilling in the Otero Mesa Wilderness[115] • Promotes idea that any oil and gas exploration is a threat to Otero Mesa[116]
Coalition for the Valle Vidal (New Mexico)	
• Prevent CBM mining in the Valle Vidal[117]	• Successfully barred oil and gas companies from the valley[118]

Defenders of Wildlife	
• Protect species and the habitats upon which they depend • Champion laws and lawmakers that protect wildlife and wild places while working against those that do them harm[119]	• Promoting a moratorium on further offshore drilling[120] • Against oil exploration and drilling in the Arctic National Wildlife Refuge[121]
Earth First!	
• Believes in biocentrism (life of the Earth comes first) and a practice of putting those beliefs into action[122]	• Organized blockade in attempt to shut down Clinch River Power Plant[123] • Opposes new coal plants in Virginia[124] • Opposes coal liquefaction[125] • Organizing a "mass mobilization" (Appalachia Rising) in Washington, D.C. to end all "mountaintop removal mining" of coal[126] • Organizes "tree sits" to oppose logging[127] • Opposes ethanol[128]
Earth Justice	
• Protect magnificent places, natural resources, and wildlife of this earth, and defend the right of all	• Impose tougher environmental restrictions on cement manufacturers[130]

people to a healthy environment[129]	• Block permits of coal-fired plants[131] • Prevent logging of old-growth trees in the Sierra Nevada[132] • Supports designating millions of acres of national forest land to prevent logging, mining, and energy development projects[133]
Earthworks	
• Works with county and municipal governments in a variety of states to enact or improve oil and gas regulations[134]	• Supports laws that govern mining on public lands in the United States[135] • Reform national mining laws[136] • Opposes creation of the following mines: Berners Bay AK (gold mine) Apache Leap AZ (copper) Pinto Creek AZ (copper) Santa Ritas AZ (copper) Uncompahgre CO (gold) Boise River ID (gold) Cabinet Mtns MT (silver/copper) Mt Tenabo NV (gold) Siskiyou OR[137]

The Environmental Defense Center (California)	
• Protects and enhances the local environment through education, advocacy, and legal action[138]	• Helped block a deep sea LNG terminal at Cabrillo port[139] • Represented groups opposed to Diablo Canyon Nuclear Power Plant • Helped defeat the proposed liquefied natural gas facility at Point Conception • Successfully stopped the expansion of the Petrochem Oil Refinery in Ventura County • Defeated ARCO's attempt to install three new platforms off the coast of UCSB • Defeated the Mobil Clearview oil project at Elwood • Convinced the National Park Service to protect natural resources at Santa Rosa Island[140]
Wild Earth Guardians (formerly Forest Guardians)	
• Protects and restores wildlife, wild rivers,	• Supports fossil fuel policy reform

and wild places in the American West[141]	• Challenges power plants and fossil fuel use[142] • Seeks halt of oil and gas permitting in Colorado[143] • Seeks to prevent logging and oil and gas "exploitation" in Carson National Forest[144] • Seeks to prevent logging, mining, and oil and gas development in 1.6 million acres of roadless areas in New Mexico via the New Mexico Water Quality Control Commission[145]
The Gifford Pinchot Task Force (Washington)	
• Works with other organizations and with coalitions to shape national legislation and Forest Service policies to protect and restore Northwest public lands[146]	• Opposes Mount St. Helens copper mine[147]
Greenpeace	
• Protect forests and the life they support • Fight to restore health to oceans • Demand U.S. climate leadership on the world	• PVC-free initiative[149] • Eliminate coal-fired power plants[150] • Cut industrial nations emissions by 80 percent

stage, work with business to implement solutions, and expose the "dirty secrets" of corporate polluters like Exxon • Eliminate toxic chemicals in rivers and lakes, air, land, and oceans • Halt the expansion of all nuclear power and shut down existing plants • Campaign for agriculture that is good for the planet and people • Support healthy food grown with the environment and farming that helps farmers to cope with climate change[148]	• Reduce carbon emissions to 1.3 tons per person per annum • Reduce electricity consumption through efficiency by 3 percent worldwide by 2050 • Phase out coal power plants[151] • Phase out oil[152] • Eliminate nuclear power[153]
Natural Resource Defense Council	
• Curb global warming and create a "Clean Energy Future" • Revive the world's oceans • Defend endangered wildlife and wild places • Protect our health by preventing pollution	• End mountaintop removal coal mining • Regulate drilling in the Marcellus Shale[155] • Prevent tar sand mining in Canada's Boreal Forest[156] • Blocked limestone mining in Florida[157]

• Ensure safe and sufficient water • Foster sustainable communities[154]	
National Wildlife Federation	
• Inspires Americans to protect wildlife for our children's future[158]	• Federal Climate Policy- seeks to put America on a path to energy independence and pass strong global warming legislation • International Climate Treaty - works with a coalition of environmental organizations around the world to push for the adoption of a new international climate change agreement • Renewable Fuels and Energy supports the expansion of clean energy sources and at the same time guide renewable energy production and transmission lines away from critical environmental areas. • Dirty Fuels supports stopping the production of crude oil extracted from tar sands and oil shale or

	produced by liquefying coal.[159]
Nuclear Information and Resource Service	
• Information and networking center for people and organizations concerned about nuclear power, radioactive waste, radiation, and sustainable energy issues.[160]	• Initiate large-scale organizing and public education campaigns on specific issues, such as preventing construction of new reactors, radioactive waste transportation, and deregulation of radioactive materials.[161]
Oil and Gas Accountability Project	
EARTHWORKS' Oil & Gas Accountability Project works with tribal, urban and rural communities to protect their homes and the environment from the devastating impacts of oil and gas development.[162]	OGAP has prepared a set of model regulations to provide county governments in New Mexico with a framework for enacting new or more stringent oil and gas regulations. A sample of the regulations includes: • County can deny permit based on any past performance anywhere in the world • Require lengthy plans including environmental and socio-economic impacts. Including county or city expansion of

	services including roads, water, sewage, housing, landfills, schools, recreation and local business. • Environmental plans must include location of all physical structures, pipelines, flowlines, and roads. Must show location of all farming and livestock activity in the vicinity. Must show location of ANY kind of water bodies (including arroyos and draws). Wildlife habitat and migration routes, as well as geologic features also must be included. • Location of all subsurface water or aquifers, the depths of the water tables and state to monitor quality of groundwater. • Must post financial guarantees of a minimum $20,000/well and $10,000/ exploration permit. • Area containing producing equipment can be no larger than 100 feet by 150 feet.

	• Site must conform to a 40 dB noise standard (this is quieter than the average home). • All pumping mechanisms must be electrically driven. • Locate all facilities at the base of slopes to provide a background of topography and natural color. • Equipment must be a uniform non-contrasting, nor-glare, non-chalking, non-reflective color tone and in a color to match the landscape and not the sky. • Roads must be surfaced with a minimum of three inches of gravel.[163]
Sierra Club	
• Explore, enjoy and protect wild places of the earth • Practice and promote the responsible use of the earth's ecosystems and resources • Educate and enlist humanity to protect and restore the quality of the natural and human environment	• Supports creation of more national monuments to restrict logging[165] • Block action to sell National Forest Land[166] • Appealed federal judge's decision to block Obama administration's six-month moratorium on deepwater drilling (2010)[167]

• Use all lawful means to carry out these objectives[164]	• Supports expanding moratoriums of offshore drilling in Virginia and Florida[168] • Convert natural hazard zones (floodplains, shorelines, landslide zones, avalanche chutes etc.) into natural areas that are free of human habitations[169] • Eliminate oil exploration and drilling in the Arctic[170] • Opposed building of Legacy Highway in Utah[171]
Theodore Roosevelt Conservation Partnership	
• Coalition of hunting, fishing and conservation organizations, labor unions, and individual grassroots partners who represent the wide spectrum of America's outdoor community. • Dedicated to the foresighted stewardship of America's landscape, helping to expand fish and wildlife habitat, and	Supports reforms to the General Mining Law of 1872, including: • Ending mining's priority status on public lands; • Recovering reasonable royalties on minerals taken from public lands and establish a fund for fish and wildlife habitat improvement to address impacts from past mining;

<table>
<tr>
<td>increasing public access to quality hunting and fishing.[172]</td>
<td>
• Ensuring that resource professionals have discretion in planning and permitting future mining to conserve public lands where high fish, water and wildlife values exist;

• Allowing reclamation incentives for Good Samaritans; and

• Prohibit patenting or sale of public lands—keep public lands in public hands.[173]

• Actively involved in a federal process to prevent the sale of leases in Colorado and other western states.

• Persuaded the BLM to cancel lease parcels in Montana, Utah and Wyoming because of sensitive fish and wildlife issues[174]

• Filed suit in U.S. District Court against the Interior Department for its mishandling of a process that cleared the way for rapidly accelerated energy development on public
</td>
</tr>
</table>

	lands in south-central Wyoming.[175] Recommends the following: • Any annual or short-term increase in federal funding for energy development should be matched by funding to deal with the consequences to fish and wildlife. • BLM and FS lands should be managed equally for multiple uses and resources • Supports a specific "Conservation Strategy" for each energy field or project, which would go beyond the NEPA-level evaluations and plans currently being completed, should be used to proactively address fish and wildlife management and needs. This Conservation Strategy should be finalized before development starts and must provide specific recommendations and actions to minimize impacts, while

	establishing plans for mitigation, detailed monitoring, and the use of adaptive management. • Recommends a change in the current leasing process that would provide for a prior assessment of impacts from lease development before leasing occurs and includes plans that balance the needs of fish and wildlife resources. • Recommends that the energy-development planning process should include science-based mitigation. This mitigation must be planned by using rigorous methods and an adaptive management process that systematically uses data from impact monitoring and evaluation to adjust development. • Supports designation of "special and unique places in the West" that should be either entirely off-limits or extremely limited to oil and gas drilling.[176]

Trout Unlimited	
• Conserve, protect and restore North America's coldwater fisheries and their watersheds[177]	• Opposes Pebble Mine (Alaska) and other large-scale mining projects[178] • Supports halt of hydraulic fracturing for gas extraction in New York[179] • Supports keeping 3.2 million acres of Wyoming "Roadless" (designed to eliminate oil and gas development in those areas[180]
San Juan Citizens Alliance (New Mexico/Colorado)	
• Organizes San Juan Basin residents to protect water and air, public lands, rural character of the area, and a unique quality of life while embracing the diversity of the region's people, economy and ecology[181]	• Supports a ban on drilling in HD mountains (Colorado)[182] • Created a *"citizens enforcement program for gas wells"* to halt increase in wells in the San Juan Mountains[183]
Save The Scenic Santa Ritas (Arizona)	
• Protects the scenic, aesthetic, recreational, environmental and wildlife values of the Santa Rita Mountains, Patagonia Mountains, Canelo Hills and San Rafael Valley through	• Opposes the opening of potential copper mines, including the Rosemont Copper Project south of Tucson, Arizona, and the Mount Emmons Mine outside of Crested Butte, Colorado.[185]

education and outreach, specifically degradation due to mining activities.[184]	
United Mountain Defense (Tennessee)	
• Protect Tennessee heritage and watersheds[186]	• Supports Appalachian Rising, a "mass mobilization" to end mountaintop removal mining.[187]
Waterkeepers Alliance	
• Protect every major watershed around the world • Collaborate with other environmental and public health organizations to serve the needs of their global network.[188]	• Hosting a contest to find the best short video telling the world that clean coal is "nothing but a dirty lie."[189] • Supports the reformation of regional, national and international energy policies by "exposing the truth behind the coal industry's dirty lies." • Working toward the protection of food systems and children from mercury released by coal-fired power plants in the United States.[190]
Western Watersheds Project	
• Protect and restore western watersheds and wildlife through education, public policy	• Supports removing livestock from public lands[192]

initiatives and litigation. • Influence and improve public land management in 8 western states; primary focus on the negative impacts of livestock grazing on 250,000,000 acres of western public lands.[191]	• Supports removing livestock from grazing on Sonoran Desert National Monument lands.[193] • Currently has case pending requesting removal of livestock from twelve million acres across five western states[194] • Won decision to remove cattle from over 800,000 acres of public land[195] • Works with other environmental groups to pass federal legislation to buy out ranchers and permanently retire public lands from grazing—all of this done with federal Funds.[196] • Opposes the creation of wind farms in Browns Bench, Idaho[197]

Endnotes

1 Sierra Club, "Home Page," Sierra Club, http://www.sierraclub.org (accessed August 2010).

2 San Juan Citizens Alliance, Blog, San Juan Citizens Alliance, http://www.sanjuancitizens.org (accessed August 2010).

3 Used with author permission from his column "On the Edge of Common Sense." For more information on Baxter Black, please check out:
Coyote Cowboy Company,
P.O. Box 2190,
Benson, AZ 85602
(800) 654-2550
www.baxterblack.com

4 Sierra Club, "Home Page, Sierra Club, http://www.sierraclub.org (accessed August 2010).

5 Defenders of Wildlife, "Wildlife Action Center, Defenders of Wildlife, http://www.savearcticrefuge.org/(accessed August 2010).

6 Mount Holyoke, "WildEarth Guardians," Mount Holyoke College, 2011 http://www.mtholyoke.edu/mwce/intern_wildearth.html (accessed August 2010).

7 San Juan Citizens Alliance, Blog, San Juan Citizens Alliance, http://www.sanjuancitizens.org (accessed August 2010).

8 CARE, "Environmental Utopia," Citizens' Alliance for Responsible Energy, http://www.responsiblenergy.org/obstruction.asp (accessed August 2010).

9 Dan Lashof, Jacqueline Wong, Laurie Johnson, "More Jobs, Less Global Warming Pollution and Greater Security for Less Than the Cost of a Postage Stamp," Natural Resources Defense Council: A Clean Energy Bargain, December 2009, http://www.nrdc.org/globalWarming/cap2.0/files/ebargain.pdf
Sierra Club, "Beyond Oil," Sierra Club, http://www.sierraclub.org/wildlands/coasts/ (accessed August 2010).
Center for Biological Diversity, "Interior Formalizes Oil-Drilling Moratorium," Center for Biological Diversity, May 31, 2010, http://www.biologicaldiversity.org/news/press_releases/2010/DrillingMoratorium-05-31-2010.html
Defenders of Wildlife, "Wildlife Action Center," Defenders of Wildlife, http://www.savearcticrefuge.org/(accessed August 2010).

10 This website points to the following website: http://www.earthfirstjournal.org/article.php?id=305 (accessed August 2010).
Defenders of Wildlife, "Wildlife Action Center, "Defenders of Wildlife, https://secure2.convio.net/dow/site/Advocacy?pagename=homepage&page=UserAction&id=713&s_einterest=C3C4(accessed August 2010).

[10] Defenders of Wildlife, "Arctic National Wildlife Refuge," Defenders of Wildlife, http://www.defenders.org/programs_and_policy/habitat_conservation/federal_lands/national_wildlife_refuges/threats/arctic/index.php (accessed August 2010). This website points to preceding website, http://www.savearcticrefuge.org/

[11] U.S. Energy Information Administration: Electricity, "Electric Power Annual: Electric Power Industry 2009: Year in Review," U.S. Energy Information Administration http://www.eia.doe.gov/cneaf/electricity/epa/epa_sum.html

[12] U.S. Energy Information Administration: Electricity, "Electric Power Annual: Electric Power Industry 2009: Year in Review," U.S. Energy Information Administration http://www.eia.doe.gov/cneaf/electricity/epa/epa_sum.html

[13] Skyler Simmons, "Full Tanks, Empty Stomachs: Ethanol and Eco-Colonialism," Earth First! Journal, http://www.earthfirstjournal.org/article.php?id=305 (accessed August 2010). Greenpeace USA, "Global Warming: Who pays the Price?" is completed, Greenpeace USA, http://members.greenpeace.org/action/start/150/ (accessed August 2010).

[14] Gena Riede, "Crazy Laws in California," *Sacramento Real Estate Voice,* February 1, 2007, http://www.sacramentorealestatevoice.com/2007/02/01/crazy-laws-in-california/

[15] J. R. McNeill, *Something New Under the Sun*, (New York, NY, W.W. Norton & Company, 2001), 310.

[16] United States History, "U.S. Population, 1790-2000: Always Growing," United States History, http://www.u-s-history.com/pages/h980.html

[17] Natural Resources Conservation Service, "Horses: A Common Sense Approach," Natural Resources Conservation Service Horse Environmental Awareness Program (HEAP), http://www.ct.nrcs.usda.gov/programs/rc&d/commonsense.html

[18] J. R. McNeill, *Something New Under the Sun*, (New York, NY, W.W. Norton & Company, 2001), 310.

[19] U.S. Census Bureau, "Population Finder," U.S. Census Bureau, http://factfinder.census.gov/servlet/SAFFPopulation?_sse=on (accessed August 2010).

[20] U.S. Census Bureau, "State & County QuickFacts," U.S. Census Bureau http://quickfacts.census.gov/qfd/states/48000.html (accessed August 2010).

[21] Paul Taylor, Executive Vice President Cary Funk, Senior Project Director April Clark, Research Associate, "Luxury or Necessity?" *PewResearchCenterPublications,* December 14, 2006, http://pewresearch.org/pubs/323/luxury-or-necessity

[22] Paul Taylor, Executive Vice President Cary Funk, Senior Project Director April Clark, Research Associate, "Luxury or Necessity?" *PewResearchCenterPublications,* December 14, 2006, http://pewresearch.org/pubs/323/luxury-or-necessity

[23] 49% of respondents said a cell phone was a necessity. All other items are higher than 50%, most well above.

[24] U.S. Energy Information Administration: Electricity, "Figure ES 1. U.S. Power Industry Net Generation," U.S. Energy Information Administration (accessed August 2010). http://www.eia.doe.gov/cneaf/electricity/epa/figes1.html (accessed August 2010).

[25] U.S. Energy Information Administration: Energy Kids: Wind, "Wind Basics," U.S. Energy Information Administration, http://www.eia.doe.gov/kids/energyfacts/sources/renewable/wind.html (accessed August 2010).

[26] U.S. Energy Information Administration, "Renewal & Alternative Fuels," EIA, http://www.eia.doe.gov/cneaf/solar.renewables/page/trends/table11.html (accessed August 2010).

[27] American Wind Energy Association, "Wind: Powering a Cleaner, Stronger America," American Wind Energy Association: Outlook 2007, http://archive.awea.org/pubs/documents/Outlook_2007.pdf

[28] Nuclear Information and Resource Service, "Home," *Nuclear Information and Resource Service* http://nirs.org/ (accessed August 2010).
Greenpeace USA , Rotating videos, Greenpeace USA http://www.greenpeace.org/usa/campaigns/nuclear/ (accessed August 2010).

[29] Keith Raether, "WWP Expands Horizons in the West, Vol. XI, No. 1, (Spring 2004) http://www.westernwatersheds.org/watmess/watmess_2004/spring_2004.pdf

[30] American Wind Energy Association, "Page Not Found," American Wind Energy Association: Outlook 2007, http://www.awea.org/pubs/documents/Outlook_2007.pdf

[31] These numbers based on a Greenpeace's goals published in "energy [r]evolution a sustainable world energy outlook" (http://www.greenpeace.org/raw/content/international/press/reports/energyrevolutionreport.pdf) which provides numbers on various sources where they envision our energy will come from by the year 2050 (pp. 46 and 47). Based on a forty-hour workweek with a windmill at 1.5 MW and solar cells that are 15% efficient.

[32] American Chemistry Council, "Page Not Found," American Chemistry Council, http://www.plasticsresource.com/s_plasticsresource/sec.asp?TRACKID=&CID=126&DID=228 (Accessed August 2010).

[33] Design News Magazine, April 30, 2007, page 49

[34] Charlyne Varkonyi Schaub, "RIP: Pink Flamingo, 1957-2006," *SunSentinel.com,* October 20, 2006, http://www.sun-sentinel.com/features/home/sfl-hg20flamingooct20,0,2739018.story?coll=sfla-features-homegarden

[35] Oil Facts, "There's a lot of life in a barrel of oil," http://docs.google.com/viewer?a=v&q=cache:C2_rkpiXIdYJ:www.scansystems.com/oil_facts.pdf+There%E2%80%99s+A+Lot+Of+LifeIn+A+Barrel+of+Oil&hl=en&gl=us&pid=bl&srcid=ADGEESgpeCzumv9V8kscexxvAD-J5WrMeCGnCH4C5w7j9oTmHOswZ690GRrrrU0pJFmZxxFWKYet4bMXwCCEQpP_CODhRHn8qABLzHFECbdyJo7-MoeMXnz0gUI26Kn-W1D2SkwPE_Fu&sig=AHIEtbTI2Fpm83toJpj-Q6FNl-Fuo4J0CA

[36] Gifford Pinchot Task Force "Progress Made in Protecting Mount St. Helens from Mining," Gifford Pinchot Task Force, http://www.gptaskforce.org/conservation/mount-st.-helens-again-threatened-by-out-of-state-mining-interests (accessed August 2010).
Save the Santa Ritas, "Home Page, Save the Santa Ritas http://www.scenicsantaritas.org/ (accessed August 2010).

[37] Robert Bryce, Power Hungry: The Myths of "Green" Energy and the Real Fuels of the Future, Public Affairs books, 93.

[38] YouTube, "John Stossel – Alternative Energy Subsidies," YouTube, Updated February 19, 2010, http://www.youtube.com/watch?v=kE23AtKMuwc

[39] CARE, "The Evolution of Fuel," Citizens' Alliance for Responsible Energy, June, 18, 2007, http://responsiblenergy.blogspot.com/2007/06/evolution-of-fuel.html

[40] Alan Portner, "First U.S. wind farm approved for Nantucket Sound over local objections," *examiner.com: Washington D.C.,* April 30, 2010, http://www.examiner.com/public-policy-in-washington-dc/first-u-s-wind-farm-approved-for-nantucket-sound-over-local-objections

[41] Brad Kuhn, "Eminent Domain Actions Planner for Sunrise Powerlink Project," *California Eminent Domain Report,* March 9, 2010, http://www.californiaeminentdomainreport.com/2010/03/articles/projects/eminent-domain-actions-planned-for-sunrise-powerlink-project/

[42] Chris Clarke, San Diego Gas & Electric blackout could have been avoided," *Desert Protective Council: Sunrise Powerlink Archive,* Sept. 8, 2011, http://dpcinc.org/index.php/site/archives/category/sunrise_powerlink/

[43] Opinion by Joseph Moore, M.D., "Power Lines and Cancer: To Move or Not to Move," *ABC News: Health*, May 30, 2007, http://abcnews.go.com/Health/CancerPreventionAndTreatment/Story?id=3214404&page=1

[44] CNN.com, "Study: Children near power lines may face higher cancer risk," *CNN.com: World,* June 3, 2006, http://www.cnn.com/2005/WORLD/europe/06/03/power.cancer/index.html

[45] Bill Opalka, "Wind Development Slowed to a Crawl," *renewablesbiz,* April 29, 2010, http://www.renewablesbiz.com/article/10/04/wind-development-slowed-crawl

[46] Deirdre Fulton, "Transmission troubles," *The Portland Phoenix,* February 11, 2009 http://portland.thephoenix.com/News/76606-Transmission-troubles/

[47] The White House, "American Jobs Act," The White House, Office of the Press Secretary, March 31, 2010, http://www.whitehouse.gov/the-press-office/remarks-president-energy-security-andrews-air-force-base-3312010

[48] American Spirit, "Oglala Sioux Tribe," http://home.comcast.net/~zebrec/Chief_Flying_Hawk.htm

[49] Joe Warron, "Cold Weather & Gas Outages," *Angel Fire, New Mexico,* February 3, 2011, http://www.angelfirenm.com/2011/02/cold-weather-gas-outages/

[50] Katie Kim "Emergency declared for energy shortage," *KRQE.com, KRQE News 13,* February 3, 2011, updated February 11, 2011, http://www.krqe.com/dpp/news/environment/emergency-declared-over-energy-shortage

[51] Free Republic, Posted by CedarDave, "Breaking: Gas Outages Reported Throughout ABQ (Up to 2,000 customers)," *The Albuquerque Journal,* February 3, 2011, http://www.freerepublic.com/focus/news/2668015/posts?page=11 (accessed August 2010).

[52] Rebecca Smith, "Texas to Probe Rolling Blackouts," *The Wall Street Journal: Business,* February 7, 2011, http://online.wsj.com/article/SB10001424052748703989504576128493806692106.html

[53] Robert Snyder, "Over-dependence on Wind Power Causes Energy Emergency in Texas," KFYO, News Talk 790 AM, February 2, 2011, http://kfyo.com/overdependence-on-wind-power-causes-energy-emergency-in-texas/

[54] Eileen O'Grady, "Loss of wind causes emergency," *Reuters: Edition U.S.*, Feb. 27, 2008, http://www.reuters.com/article/2008/02/28/us-utilities-ercot-wind-idUSN2749522920080228

[55] Kate Galbraith, "An Interview With the CEO of the Texas Grid," *The Texas Tribune,* February 4, 2011, http://www.texastribune.org/texas-energy/energy/an-interview-with-the-ceo-of-the-texas-grid/

[56] responsible energy, "The Obama Energy Plan. What Can We Expect?" YouTube, updated Jan. 12, 2009, http://www.youtube.com/watch?v=oEmXUqbPJ18&feature=player_embedded (accessed August 2010).

[57] Tom Doggett, "Obama to announce loan help for nuclear power," *Reuters: Edition U.S.*, Feb. 15, 2010, http://www.reuters.com/article/2010/02/16/us-usa-nuclear-loan-idUSTRE61F0BR20100216

[58] Tennessee Valley Authority, "Watts Bar Nuclear Plant," Power System, Tennessee Valley Authority, http://www.tva.gov/power/nuclear/wattsbar.htm (accessed August 2010).

[59] U. S. Wind Energy Projects, "National Total Power Capacities," U. S. Wind Energy Projects, (As of September, 30, 2010)., http://archive.awea.org/projects/(accessed August 2010).

[60] U.S. Department of Energy: Energy Efficiency & Renewable Energy, EERE State Activities & Partnership, "States with Renewable Portfolio Standards," U.S. Department of Energy, http://apps1.eere.energy.gov/states/maps/renewable_portfolio_states.cfm (accessed August 2010).

[61] U.S. Department of Energy: Energy Efficiency & Renewable Energy, EERE State Activities & Partnership, "States with Renewable Portfolio Standards," U.S. Department of Energy, http://apps1.eere.energy.gov/states/maps/renewable_portfolio_states.cfm (accessed August 2010).

[62] American Chemistry Council, "History of Polymer and Plastics for Students," American Chemistry Council, http://www.americanchemistry.com/hops/intro_to_plastics/students.html

[63] Plastics Make it Possible, "Plastics in Your Life," Plastics Make it Possible Health and Safety, http://plasticsmakeitpossible.com/category/plastics-in-your-life/health-safety/ http://www.americanplasticscouncil.org/benefits/in_your_life/pop/power_plastics9-99.html

[64] American Chemistry Council, "Packaging and consumer products," American Chemistry Council, http://vwww.polyurethane.org/s_plastics/sec_content.asp?CID=1078&DID=4232&dowhat=&css=print

[65] Drake Bennett, "The locavore's dilemma: Sometimes buying local food helps in the battle against climate change. Sometimes it doesn't. And sometimes, it's just too confusing to decide." *The Boston Globe,* July 22, 2007. http://www.boston.com/news/globe/ideas/articles/2007/07/22/the_localvores_dilemma/

[66] WildEarth Guardians: A Force for Nature, "Features," WildEarth Guardians http://www.wildearthguardians.org/ (accessed August 2010).

[67] Sierra Club, " Resilient Habitats: Ecosystems," Sierra Club, http://www.sierraclub.org/habitat/ecosystems/bighorn.aspx Earth First!, " The View from the Trees - spring & summer in Cascadia, USA," Earth First! https://earthfirst.org.uk/actionreports/node/22852 Keith Raether, "WWP Expands Horizons in the West, Vol. XI, No. 1, (Spring 2004) http://www.westernwatersheds.org/watmess/watmess_2004/spring_2004.pdf

[68] Mark Matthews, "Finger Pointing Begins Over Tahoe Fire, ABC, KGO-TV San Francisco, CA, June, 25, 2007, http://abclocal.go.com/kgo/story?section=politics&id=5421815

[69] Brendan I. Koerner, "Dirty Burns," Home, The Green Lantern: Illuminating Answers to Environmental Questions, *Slate,* Oct. 16, 2007, http://www.slate.com/id/2175938/

[70] Natural Resources Defense Council: The Earth's Best Defense, "Federal Judge Blocks Huge Florida Mining Project That Threatens Miami Water, Everglades," News: Media Center, *Natural Resources Defense Council: The Earth's Best Defense*, March 22, 2006 http://nrdc.org/media/pressreleases/060322.asp

[71] Wikipedia, "Flying Hawk," Wikipedia http://en.wikipedia.org/wiki/Flying_Hawk

[72] Peter Schweizer, "Gore isn't quite as green as he's led the world to believe," Editorial/Opinion, *USA Today,* December, 7, 2006, http://www.usatoday.com/news/opinion/editorials/2006-08-09-gore-green_x.htm

[73] Answers.com, "Chemical Industry," Answers.com, http://www.answers.com/topic/chemical-industry

[74] Centers for Disease Control and Prevention, "Parasites," Centers for Disease Control and Prevention, http://www.cdc.gov/ncidod/dpd/parasites/giardiasis/factsht_giardia.htm#transmission (accessed August 2010).

[75] Eric R. Pianka , "The Vanishing Book of Life on Earth" http://uts.cc.utexas.edu/~varanus/Vanishing.Book.text.pdf, Sarah-Kate Templeton, "Having large families 'is an eco-crime'," The Times, May 6, 2007 http://women.timesonline.co.uk/tol/life_and_style/women/families/article1752235.ece

[76] United States Environmental Protection Agency, "Air Emissions Summary Through 2005," Air Trends, United States Environmental Protection Agency, http://www.epa.gov/air/airtrends/2006/emissions_summary_2005.html (accessed August 2010).

[77] Fly Rod and Reel Magazine, April 2007, p. 20

[78] Keith Raether, "WWP Expands Horizons in the West, Vol. XI, No. 1, (Spring 2004) http://www.westernwatersheds.org/watmess/watmess_2004/spring_2004.pdf

[79] The Santa Fe New Mexican, July 17, 1.

[80] Ron Arnold, "Left Tracking Library" and "The Green Tracking Library," Project of the Center for the Defense of Free Enterprise, http://www.undueinfluence.com/ (accessed August 2010).

[81] George F. Will, "March of the Polar Bears," *The Washington Post,* May 22, 2208, http://www.washingtonpost.com/wp-dyn/content/article/2008/05/21/AR2008052102428.html

[82] Fred L, Smith, Jr. "Eco-Fascism Going Global," op-ed in USA Next, *Competitive Enterprise Institute,* Sept. 13, 2004. http://cei.org/op-eds-and-articles/eco-fascism-going-global

[83] "Joseph Goebbels: The Poison Dwarf," Holocaust Education & Archive Research Team. http://www.holocaustresearchproject.org/holoprelude/goebbels.html (accessed August 2010).

[84] Edward Bernays, "History Is a Weapon: Propaganda (1928)" http://www.historyisaweapon.com/defcon1/bernprop.html (accessed August 2010).

[85] Paul Driessen, "Wanted: Ethics for activists," *Eco Imperialism,* February 2005 http://www.eco-imperialism.com/content/printview.php3?id=113

[86] Source: Congressional Record V. 148, Pt. 4, April 11, 2002 to April 24, 2002 http://books.google.com/books?id=iOgfSDKecCcC&pg=PA4860&lpg=PA4860

[87] Jason Mraz: From Grammys to going green, http://www.cbs8.com/story/12078278/jason-mraz-from-grammys-to-going-green

[88] Sandra Sanchez, "Simon Carter/EcoGir, *Orglamic,* March 8, 2010. http://orglamic.com/tag/jason-mraz/

[89] Reviewed by Joshua Zyber, "The X-Files: I Want to Believe," *Hi-Def Digest,* December 24, 2008 http://bluray.highdefdigest.com/1625/xfiles_iwanttobelieve.html

[90] Wynne Parry, "Global Warming: Dire Prediction for the Year 3000," *Live Science,* January 12, 2011. http://www.livescience.com/9281-global-warming-dire-prediction-year-3000.html

[91] Jeffrey Kluger, "Earth at the Tipping Point: Global Warming Heats Up," Time Frames, Environment, *Time Magazine,* March 26, 2006. http://www.time.com/time/magazine/article/0,9171,1176980-6,00.html

[92] EDF, "Tips to fight global warming," Environmental Defense Fund, http://fightglobalwarming.com/page.cfm?tagID=137 (accessed August 2010).

[93] platformbreathe, "Support Clean Energy. Now," platformbreathe, 2008 http://www.platformbreathe.org/page2/page2.html

[94] Northwest Mining Association, "Solar Power Need American Metals & Minerals," Northwest Mining Association, http://www.nwma.org/pdf/Solar_Power_Needs_American_Minerals%20_2_.pdf (accessed August 2010).

[95] Robert Bryce, Power Hungry, The Myth of "Green Energy and the Real Fuels of the Future (New York New York: Public Affairs Books, an imprint of Perseus Books Group, 2010). page 93.

[96] The Environmental Source, "The Clean Air Act," *Competitive Enterprise Institute,* http://cei.org/pdf/2314.pdf (accessed August 2010).

[97] Senator Jim Webb, Senator Webb: Congress Should End Costly Ethanol Subsidies, Press Releases, November 30, 2010 http://webb.senate.gov/newsroom/pressreleases/11-30-2010-02.cfm

[98] Review & Outlook, "Al Gore's Ethanol Epiphany," *The Wall Street Journal,* November 27, 2010. http://online.wsj.com/article/SB10001424052748703572404575634753486416076.html

[99] Director's Blog, "Ethanol, Food Prices, and Greenhouse-Gas Emissions," *Congressional Budget Office,* April 8, 2009, http://cboblog.cbo.gov/?p=228

[100] Rich Lowry, "The Ethanol Idiocy that Will Not Die," *National Review Online,* King Features Syndicate, 2010. http://www.nationalreview.com/articles/255221/ethanol-idiocy-will-not-die-rich-lowry

[101] "City Leaves Green to Get Green: Albuquerque City Vehicles Moving Away From Alternate Fuel," *KOAT TV,* Posted May 2, 2010, Updated May 3, 2010. http://www.koat.com/money/23426132/detail.html

[102] Ted Williams, "America's corn-based ethanol program carries high costs in fish, wildlife and tax dollars," Under the Influence of Ethanol, *Fly Rod & Reel Online,* April 2007 http:*//www.flyrodreel.com/magazine/2007/april/under-influence-ethanol*

[103] John Carey, Posted by Dan Beucke, "House Passes Carbon Cap-and-Trade Bill," Money & Politics blog, *Bloomberg Businessweek, June 26, 2009* http://www.businessweek.com/blogs/money_politics/archives/2009/06/house_passes_ca.html

[104] Caribou Clean Water Partnership, "About the Partnership," Caribou Clean Water Partnership, http://www.caribou cleanwater.org/about.html (accessed August 2010).

[105] Caribou Clean Water Partnership, "Streamline Newsletter," *Caribou Clean Water Partnership,* May 2010, http://www.caribou cleanwater.org/streamline/

[106] Center for Biological Diversity, "About the Center, Our Story," Center for Biological Diversity, http://www.biologicaldiversity.org/about/story/index.html (accessed August 2010).

[107] Center for Biological Diversity, "Public Lands," Center for Biological Diversity, http://www.biologicaldiversity.org/programs/public_lands/index.html (accessed August 2010).

[108] Center for Biological Diversity, "About the Center: Our Story," Center for Biological Diversity, http://www.biologicaldiversity.org/about/story/index.html (accessed August 2010).

[109] Center for Biological Diversity, "Saving the Razorback Sucker," Center for Biological Diversity, http://www.biologicaldiversity.org/species/fish/razorback_sucker/index.html (accessed August 2010).

[110] Clark Fork Coalition, "Home," Clark Fork Coalition, http://www.clarkfork.org/ (accessed August 2010).

[111] Clark Fork Coalition, "404: Not Found," Clark Fork Coalition, http://www.clarkfork.org/take-action/take-action-alerts-page.html (Accessed August 2010).

[112] Clark Fork Coalition, "Rock Creek Mine," Clark Fork Coalition, http://www.clarkfork.org/water-watch/rock-creek-mine.html (accessed August 2010).

[113] Environmental News Service, "Court Blocks Mine in Montana's Cabinet Mountains Wilderness," Environmental News Service, March 29, 2010 http://www.ens-newswire.com/ens/mar2010/2010-03-29-092.html

[114] Otero Mesa: American's Wildest Grassland, "Write a Letter to Secretary of the Interior Salazar," Otero Mesa: American's Wildest Grassland, http://www.oteromesa.org/featured-articles/write-a-letter-to-secretary-of-the-interior-salazar/ (accessed August 2010).

[115] Otero Mesa: American's Wildest Grassland, "The Campaign to Protect Otero Mesa," Otero Mesa: American's Wildest Grassland, http://www.oteromesa.org/the-campaign-to-protect-otero-mesa/ (accessed August 2010).

[116] Otero Mesa: American's Wildest Grassland, "Oil and Gas," Otero Mesa: American's Wildest Grassland, http://www.oteromesa.org/threats/oil-and-gas/ (accessed August 2010).

[117] Coalition for the Valle Vidal, "Home," Coalition for the Valle Vidal, http://www.vallevidal.org/ (accessed August 2010).

[118] Coalition for the Valle Vidal, "Overview," Coalition for the Valle Vidal http://www.vallevidal.org/overview.php (accessed August 2010).

[119] Defenders of Wildlife, "About Us," and "Frequently Asked Questions," Defenders of Wildlife, http://www.defenders.org/about_us/faqs/general.php (accessed August 2010).

[120] Defenders of Wildlife, "Disaster in the Gulf," Defenders of Wildlife, http://www.defenders.org/programs_and_policy/offshore_drilling/gulf_oil_disaster/index.php(accessed August 2010).

[121] Defenders of Wildlife, "Arctic National Wildlife Refuge," Defenders of Wildlife, http://www.defenders.org/programs_and_policy/habitat_conservation/federal lands/national_wildlife_refuges/threats/arctic/index.php (accessed August 2010).

[122] Earth First! Journal, "About Earth First!" Earth First! Worldwide, http://earthfirst.org/about.htm (accessed August 2010).

[123] Earth First! Journal, "American Electric Power's Clinch River Coal Power Plant," Earth First! Journal, http://www.earthfirstjournal.org/article.php?id=275 (accessed August 2010).

[124] "Tell Dominion Power No New Coal Plants in Virginia," *Mountain Justice*, June 21, 2007, 5:42 pm http://www.mountainjusticesummer.org/events.php?id=70

[125] Mountain Justice, "Monday, June 16: Stop Coal to Oil Boondoggle in Pikeville, KY." Mountain Justice, http://www.mountainjusticesummer.org/events.php?id=71 (accessed August 2010).

[126] Appalachia Rising, "Home," Appalachia Rising, http://appalachiarising.org/ (accessed August 2010).

[127] Treesit/blogspot "One Year Later, RMCP Investigator Questions Tree Sit Raid," Treesit/blogspot http://treesit.blogspot.com/ (accessed August 2010).

[128] Tinder, "Resistance Heats up in the Biscuit," Earth First! Journal http://www.earthfirstjournal.org/article.php?id=227 (accessed August 2010). Skyler Simmons, "Full Tanks, Empty Stomachs: Ethanol and Eco-Colonialism," Earth First! Journal, http://www.earthfirstjournal.org/article.php?id=305 (accessed August 2010).

[129] EarthJustice, "About Us," EarthJustice, http://earthjustice.org/about (accessed August 2010).

[130] EarthJustice, "Air Toxics," EarthJustice, http://www.earthjustice.org/features/air-toxics (accessed August 2010).

[131] EarthJustice, "Our Cases" EarthJustice Our Work, 2007 http://earthjustice.org/our_work/cases/2007/page.jsp?itemID=30390639

[132] EarthJustice, "The Creation of National Forest Roadless Area Conservation Policy, EarthJustice: Timeline of the Roadless Rule, 2009-1998 http://earthjustice.org/features/timeline-of-the-roadless-rule

[133] EarthJustice, "Protecting Our Sierra Nevada," EarthJustice, http://www.earthjustice.org/sites/default/files/library/factsheets/SNF_Fact_Sheet.pdf (accessed August 2010).

[134] Earthworks, "Home," Earthworks, http://www.earthworksaction.org/home.cfm (accessed August 2010).

[135] Earthworks, "The Way Forward," Earthworks, http://www.earthworksaction.org/TheWayForward.cfm (accessed August 2010).

[136] Earthworks, "1872 Mining Law," Earthworks, http://www.earthworksaction.org/1872.cfm (accessed August 2010).

[137] Earthworks, "Special Places," Earthworks, http://www.earthworksaction.org/SpecialPlaces.cfm (accessed August 2010).

[138] Environmental Defense Center, "Home," Environmental Defense Center http://www.edcnet.org/ (accessed August 2010).

[139] Matthew Singer, "Cabrillo Port runs out of gas," April 12, 2007, *VCReporter Newsletter* http://www.vcreporter.com/cms/story/detail/?id=4523&IssueNum=119

[140] Environmental Defense Center, "Case History," Environmental Defense Center http://www.edcnet.org/learn/Case_History.html (accessed August 2010).

[141] WildEarth Guardians, "Home," WildEarth Guardians, http://www.wildearthguardians.org/ (accessed August 2010).

[142] WildEarth Guardians, "History" WildEarth Guardians, About Us, http://www.wildearthguardians.org/site/PageServer?pagename=about_history (accessed August 2010).

[143] WildEarth Guardians, "Clean Air at Risk from Expanded Oil and Gas Drilling Along Colorado's Front Range," WildEarth Guardians Newsroom, January 19, 2011, http://www.wildearthguardians.org/site/News2?page=NewsArticle&id=6493&news_iv_ctrl=1230

[144] WildEarth Guardians, " Carson National Forest Tops List of Threatened Forests: New Report Reveals Administration's Push for More Oil & Gas Drilling," WildEarth Guardians Newsroom, October 11, 2005, http://www.wildearthguardians.org/site/News2?page=NewsArticle&id=6004

[145] WildEarth Guardians, "First Petition to Nominate Roadless Forest Streams As Outstanding National Resource Waters Submitted," WildEarth Guardians Newsroom, *October 25, 2006,* http://www.wildearthguardians.org/site/News2?page=NewsArticle&id=5787

[146] Gifford Pinchot Task Force, "Policy," Gifford Pinchot Task Force http://www.gptaskforce.org/policy (accessed August 2010).

[147] Gifford Pinchot Task Force "Progress Made in Protecting Mount St. Helens from Mining," Gifford Pinchot Task Force http://www.gptaskforce.org/conservation/mount-st.-helens-again-threatened-by-out-of-state-mining-interests (accessed August 2010).

[148] Greenpeace USA, "Register and donate," Greenpeace USA http://www.greenpeace.org/usa/en/campaigns/ (accessed August 2010).

[149] Greenpeace USA "Go PVC-Free" Greenpeace USA http://www.greenpeace.org/usa/campaigns/toxics/go-pvc-free/ (accessed August 2010).

[150] Greenpeace, Clean Energy Future, "The Dirty Dozen, Twelve new coal-fired power stations are on the table for Australia," Greenpeace Australia/Pacific http://www.greenpeace.org.au/climate/GI-dirtydozen.html (accessed August 2010).

[151] Greenpeace, "How to Save the Climate: Join the Energy (R)evolution," Greenpeace Media GmbH, Hamburg/ Editor-in-chief: Jochen Schildt (accessed August 2010). http://www.greenpeace.org/raw/content/international/press/reports/how-to-save-the-climate-pers.pdf

[152] Greenpeace USA, "Ruptured Oil Tanker Sinks off Coast of Spain: Spill Feared to Be Larger than Exxon Valdez," Greenpeace USA, Nov. 20, 2002 http://www.greenpeace.org/usa/en/news-and-blogs/news/ruptured-oil-tanker-sinks-off/

[153] Greenpeace USA, "New Yorkers Rally to Shut Down Indian Point, " Greenpeace USA What We Do, http://www.greenpeace.org/usa/campaigns/nuclear/

[154] Natural Resources Defense Council, "About Us Page," Natural Resources Defense Council, http://www.nrdc.org/about/ (accessed August 2010).

[155] Natural Resources Defense Council, Policy Solutions, Natural Resources Defense Council, http://www.nrdc.org/policy/ (accessed August 2010).

[156] Natural Resources Defense Council, Land Facts, "Strip Mining for Oil in Endangered Forests," Natural Resources Defense Council, June 2006 http://www.nrdc.org/media/docs/060607a.pdf

[157] Natural Resources Defense Council, "Federal Judge Blocks Huge Florida Mining Project that Threatens Miami Water, Everglades," Press Release, Natural Resources Defense Council http://www.nrdc.org/media/pressreleases/060322.asp (accessed August 2010).

[158] National Wildlife Federation, "Homepage," National Wildlife Federation, http://www.nwf.org/ (accessed September 2011)

[159] National Wildlife Federation, "Policy Solutions to Global Warming: Reducing Harmful Drilling and Mining," National Wildlife Federation Global Warming Policy, http://www.nwf.org/Global-Warming/Policy-Solutions.aspx (accessed September 2011)

[160] Nuclear Information and Resource Service, "Home," *Nuclear Information and Resource Service* http://nirs.org/ (accessed September 2011)

[161] Nuclear Information and Resource Service, "About NIRS," Nuclear Information and Resource Service, http://www.nirs.org/about/nirs.htm (accessed September 2011)

[162] Earthworks, "Oil and Gas Accountability Project," Earthworks http://www.earthworksaction.org/oil_and_gas.cfm (accessed September 2011)

[163] Thomas F. Darin and Travis Stills, "Preserving Our Public Lands; a citizen's guide to understanding and participating in oil and gas decisions affecting our public lands, Sponsored by: Land & Water Fund of the Rockies, Oil & Gas Accountability Project, and Wyoming Outdoor Council, 2002, 1993 by the Land and Water Fund of the Rockies. http://www.earthworksaction.org/pubs/Preserving%20our%20public%20lands.pdf

[164] Sierra Club, Sierra Magazine, Sierra Club, http://sierraclub.org/ (accessed September 2011)

[165] Maggie Bridgeman, "Bill would limit president's power to create national monuments," *McClatchy Newspapers,* July 1, 2010 http://www.mcclatchydc.com/2010/07/01/96885/bill-would-limit-presidents-power.html

[166] Sierra Club, "Conservation Corner," Sierra Club, http://southcarolina.sierraclub.org/lunz/conservation.html (accessed August 2010).

[167] Sierra Club, "Sierra Club Takes Legal Action as Oil Disaster Continues to Devastate Gulf Coast," Sierra Club, http://www.sierraclub.org/environmentallaw/lawsuits/0442.aspx (accessed August 2010).

[168] Michael Brune, "The BP Oil Disaster," Sierra Club, http://action.sierraclub.org/site/PageNavigator/adv_oilspill (accessed August 2010).

[169] Sierra Club, "Stop Polluters," Registration Page, Sierra Club, http://www.sierraclub.org/communities/ (accessed August 2010).

[170] Sierra Club, "Chill the Drills: Protect America's Arctic!," Sierra Club, http://www.sierraclub.org/arctic/ (accessed August 2010).

[171] Sierra Club, "Stop Sprawl," Sierra Club, http://www.sierraclub.org/sprawl/report00/roads.asp (accessed August 2010).

[172] Theodore Roosevelt Conservation Partnership, Theodore Roosevelt Conservation Partnership About page, Theodore Roosevelt Conservation Partnership, http://www.trcp.org/about.html (accessed August 2010)..

[173] Theodore Roosevelt Conservation Partnership, "Mining Issues," Theodore Roosevelt Conservation Partnership, http://www.trcp.org/issues/mining.html (accessed August 2010).

[174] Theodore Roosevelt Conservation Partnership, "Page not found," Theodore Roosevelt Conservation Partnership, http://www.trcp.org/newsroom/pressreleases/17-pressreleases/119-pr-2007-10-24.html (accessed August 2010).

[175] Theodore Roosevelt Conservation Partnership, "Energy Issues," Theodore Roosevelt Conservation Partnership, http://www.trcp.org/issues/energy.html (accessed August 2010).

[176] Theodore Roosevelt Conservation Partnership, "Funding Accountability Coordination Transparency Science," Theodore Roosevelt Conservation Partnership Issues Energy, http://www.trcp.org/issues/energy/facts-for-fish-and-wildlife

[177] Trout Unlimited, "About Us Page," Trout Unlimited, http://www.tu.org/about-us (accessed September 2011)

[178] Trout Unlimited , "Save Bristol Bay," Trout Unlimited Special Projects, http://www.tu.org/about-us/special-projects (accessed September 2011)

[179] Trout Unlimited, "Trout Unlimited and Theodore Gordon Flyfishers Praise New York State Senate for Passing Bill Delaying Gas Drilling," *Trout Unlimited,* Press release, August 4, 2010 http://www.tu.org/press_releases/2010/trout-unlimited-and-theodore-gordon-flyfishers-praise-new-york-state-senate-for-

[180] Wyoming Wilderness, 2008 Newsletter no longer available, Wyoming Wilderness, http://www.wildwyo.org/WWAnewsletters/WWA2008Newsletter.pdf (accessed September 2011)

[181] San Juan Citizens Alliance, "Clean Water Network," San Juan Citizens Alliance, http://www.cleanwaternetwork.org/members/orgs/san-juan-citizens-alliance

[182] The HD Mountains: Southwest Colorado's Threatened Jewel, "Coalition Files suit Against HD Mountains Decision," Save the HD Mountains, January 23, 2008, http://www.savehdmountains.org/

[183] Mark Pearson, http://www.cleanwaternetwork.org/members/orgs/san-juan-citizens-alliance Canyons of Ancients Seismic Suit Settled," *San Juan Citizens News*, March 2003 http://www.sanjuancitizens.org/otherpages/Newsletters/SJCANews_Mar03.pdf

[184] Save the Santa Ritas, "Home Page, Save the Santa Ritas http://www.scenicsantaritas.org/ (accessed September 2011)

[185] Save the Santa Ritas, Rosemont Mine, "Overview," Save the Santa Ritas, http://www.scenicsantaritas.org/rosemont.asp (accessed September 2011)

[186] United Mountain Defense, "Home Page," United Mountain Defense, Protecting Tennessee's Watersheds, Air, Mountains, and People Blog, http://www.unitedmountaindefense.org/ (accessed September 2011)

[187] Appalachia Rising blog, "Home Page," Appalachia Rising, http://appalachiarising.org/ (accessed August 2010).

[188] Waterkeeper Alliance, "Our Mission," Waterkeeper Alliance, http://www.waterkeeper.org/ht/d/sp/i/187/pid/187 (accessed August 2010)

[189] Waterkeeper Alliance, "The Watering Hole Blog," Waterkeeper Alliance, http://www.waterkeeper.org/ht/d/Contents/contenttype_id/21/pid/13920 (accessed August 2010).

[190] Waterkeeper Alliance, "Clean Energy Campaign," Waterkeeper Alliance, http://waterkeeper.org/ht/display/ContentDetails/i/716/pid/196 (accessed August 2010).

[191] Western Watersheds Project, "About Western Watersheds Project," Western Watersheds Project, (http://www.westernwatersheds.org/about (accessed August 2010).

[192] Katie Fite, "Biodiversity Bulletin," *Watersheds Messenger*, Vol. XI, No. 1 (Spring 2004) https://www.westernwatersheds.org/watmess/watmess_2004/2004html_spring/article7.htm

[193] Greta Anderson, "Not a Good Place for Cows," *Watersheds Messenger*, Vol. XV, No. 2 (Fall 2008) http://www.westernwatersheds.org/watmess/watmess_2008/WWPMessenger10-0.pdf

[194] Laird Lucas, WWP Litigation, "WWP Legal Efforts are Winning!" Western Watersheds Project, http://www.westernwatersheds.org/legal/legal.html

[195] Michelle Dunlop, "Western Watersheds Project Wins A Major Federal Court Victory Ordering The Immediate Removal Of Cattle From Over 800,000 Acres Of Public Lands In The Jarbidge Field Office Of The BLM!" *WWP Online Messenger* #99, (August 4, 2005) http://www.westernwatersheds.org/news_media/newsmedia_2005/wwp99_newsmedia.html

[196] Jon Marvel, National Public Lands Gazing Campaign, "WWP & the National Public Lands Grazing Campaign," *Western Watersheds Project* http://www.westernwatersheds.org/NPLGC/nplgc.html

[197] Katie Fite, "Biodiversity Bulletin," *Watersheds Messenger, Vol. XI, No.1, (Spring 2004)* https://www.westernwatersheds.org/watmess/watmess_2004/2004html_spring/article7.htm

Intermedia Publishing Group

Publishing That Works For You

Do you need a speaker?

Whether you want to purchase bulk copies of *Energy Freedom* or buy another book for a friend, get it now at: www.imprbooks.com.

Do you want Marita Littauer Noon to speak to your group or event? Then contact Larry Davis at: (623) 337-8710 or email: ldavis@intermediapr.com or use the contact form at: www.intermediapr.com.

For your publishing needs, contact Terry Whalin, Publisher, at Intermedia Publishing Group, (623) 337-8710 or email: twhalin@intermediapub.com or use the contact form at: www.intermediapub.com.

"Marita Noon speaks to the heart of the energy crisis facing America today, namely the apparent wish of so many to turn the lights out. She is a vital voice of sanity and common sense in the energy conversation. This book could not be more timely."

Ann McElhinney, Journalist and documentary filmmaker
Director, *Mine Your Own Business* and *Not Evil Just Wrong*

"I have mentored many people learning the politics of environmentalism and how to confront its excesses. Marita Noon has been the most rewarding, not just in her quick grasp of the issues and her uncanny ability to see what needs to be done, but also in her intelligent and swift action. Since we began to work together in 2007, I've watched her grow from an eager learner to an in-demand expert who can address energy issues in outstanding articles and widely published op-eds, as well as drawing fascinated viewers and listeners on TV and radio. She is magic in front of an audience. I could not be more proud of her and her new book ***Energy Freedom***."

Ron Arnold, Executive Vice President, Center for the
Defense of Free Enterprise, Columnist, *Washington Examiner*

There is an incontrovertible historical relationship between the preservation of freedom in America and American's access to increasingly available fixed and portable energy supplies. With the advance of technological threats to freedom since Independence, accelerating energy production and conversion efficiencies have provided the economic growth and resources critical to national defense and to our internal freedom to enjoy "life, liberty, and the pursuit of happiness". Marita Noon's *Energy Freedom* documents both the role energy plays in the exercise of American freedom and the efforts by some to subvert that freedom through restricting its supply. Voters and consumers should study *Energy Freedom* carefully, and it should be required reading for all who seek the public trust whether in enterprise or politics.

Harrison H. Schmitt – Geologist, Apollo 17 Astronaut,
Former U.S. Senator, and author of *Return to the Moon* and
the "America's Uncommon Sense" web site

"We are in a struggle in this country. On the one side are those who believe that freedom, and the resulting individual productivity, have been the keys to this nation's success. On the other side are those who would allow sentiment to stifle freedom, regardless of the adverse impact on productivity and the wellbeing of the nation. And, when bad consequences flow from this latter approach, its proponents exude a mirage of excuses to cloud the reasons for their failures. One who pushes back against this latter approach is Marita Noon. She as been a voice for freedom and productivity. I, for one, appreciate her work in this area."

Harvey E. Yates, Jr.,
New Mexico State Republican Party Chairman, 2009-2010

"Environmental groups are fighting to destroy American exceptionalism, American jobs, and American energy. Fortunately there are a few brave people fighting to keep them from diminishing American industry and American prosperity. Marita Noon is one of them. She is on the front lines of the battle."

Steve Pearce, United States Congress, NM 2

"Even people who have no idea where the electricity that lights their home or gas that cooks their meals comes from will find ***Energy Freedom*** a clear, understandable primer on why we all have to care about energy. Marita Noon lays out all the obstacles blocking a balanced energy approach that can allow us to run a prosperous economy, protect the environment and drive crazy the overseas dictators who think they have us over a barrel of their oil."

John Fund, Columnist, *Wall Street Journal*

"Marita Noon is a true freedom fighter, making a welcome impact in the debate over America's energy policy and future."

Christopher C. Horner
Senior Fellow Competitive Enterprise Institute
Author, ***Red Hot Lies: How Global Warming Alarmists Use Threats, Fraud and Deception to Keep You Misinformed*** and ***The Politically Incorrect Guide to Global Warming and Environmentalism***

"The Sierra Club has never seen an energy project they liked. As you'll find out when you read ***Energy Freedom***, this statement is true for the environmental movement as a whole. Marita Noon shows, that not only do environmentalists not like any energy project, they actually aim to shut them down—leaving a diminished America in a reduced energy environment. This, despite the fact that America is rich with natural resources. Marita always has a fresh take on energy issues that has made her a popular contributor to the *Energy Tribune*."

Prof. Michael J. Economides
Cullen College of Engineering, University of Houston
Editor-in-Chief *Energy Tribune*

"Marita Noon is a real voice, making a big difference. She is meeting the ever increasing energy opportunity for free markets and societal progress."

Robert L. Bradley, Jr.
CEO and founder of the Institute for Energy Research
Author of six books including ***Capitalism at Work: Business, Government, and Energy*** and ***Energy: The Master Resource***